TRIANGULAR NORM-BASED MEASURES AND GAMES WITH FUZZY COALITIONS

THEORY AND DECISION LIBRARY

General Editors: W. Leinfellner (*Vienna*) and G. Eberlein (*Munich*)

Series A: Philosophy and Methodology of the Social Sciences

Series B: Mathematical and Statistical Methods

Series C: Game Theory, Mathematical Programming and Operations Research

Series D: System Theory, Knowledge Engineering and Problem Solving

SERIES C: GAME THEORY, MATHEMATICAL PROGRAMMING AND OPERATIONS RESEARCH

VOLUME 10

Scope: Particular attention is paid in this series to game theory and operations research, their formal aspects and their applications to economic, political and social sciences as well as to socio-biology. It will encourage high standards in the application of game-theoretical methods to individual and social decision making.

The titles published in this series are listed at the end of this volume.

TRIANGULAR NORM-BASED MEASURES AND GAMES WITH FUZZY COALITIONS

by

DAN BUTNARIU
Department of Mathematics and Computer Science, University of Haifa, Israel

and

ERICH PETER KLEMENT
Institute of Mathematics, Johannes Kepler University, Linz, Austria

KLUWER ACADEMIC PUBLISHERS
DORDRECHT / BOSTON / LONDON

Library of Congress Cataloging-in-Publication Data

Butnariu, Dan.
Triangular norm-based measures and games with fuzzy coalitons / by Dan Butnariu and Erich Peter Klement.
p. cm. -- (Theory and decision library. Series C, Game theory, mathematical programming, and operations research ; v. 10)
Includes bibliographical references and index.
ISBN 0-7923-2369-6 (acid-free)
1. Game theory. 2. Fuzzy sets. I. Klement, E. P. (Erich Peter)
II. Title. III. Series.
QA269.B88 1993
519.3--dc20 93-24820

ISBN 0-7923-2369-6

Published by Kluwer Academic Publishers,
P.O. Box 17, 3300 AA Dordrecht, The Netherlands.

Kluwer Academic Publishers incorporates
the publishing programmes of
D. Reidel, Martinus Nijhoff, Dr W. Junk and MTP Press.

Sold and distributed in the U.S.A. and Canada
by Kluwer Academic Publishers,
101 Philip Drive, Norwell, MA 02061, U.S.A.

In all other countries, sold and distributed
by Kluwer Academic Publishers Group,
P.O. Box 322, 3300 AH Dordrecht, The Netherlands.

Printed on acid-free paper

Printed in the Netherlands

To Hadassah and Sieglinde

Contents

Acknowledgment

We are very grateful to *Robert J. Aumann, Irinel Dragan, Lloyd S. Shapley*, and *Lotfi A. Zadeh* for repeatedly giving us the opportunity to discuss various aspects of the material presented in this work. We greatly benefitted from their deep insights into different fields and also from their personal support.

We are indebted to *Zvi Artstein, Radko Mesiar, Roland Takacs*, and *Anton Wakolbinger* whose solutions to some mathematical problems led to improvements in our own results.

We greatly appreciate the comments of *Alain Chateauneuf, Ulrich Höhle, Vilém Novák, Endre Pap*, and *Siegfried Weber* who kindly agreed to read large parts of the manuscript and whose comments helped us to improve earlier versions of it.

We gratefully acknowledge the following institutions and individuals: the Austrian *Bundesministerium für Wissenschaft und Forschung*, Vienna, the *Linzer Hochschulfonds*, the *Österreichische Mathematische Gesellschaft*, our colleagues in the *Institut für Mathematik* of the *Johannes Kepler Universität Linz*, Austria (in particular *Peter Weiß*), in the *Institut für Systemtheorie* of the *Johannes Kepler Universität Linz*, Austria (especially *Franz Pichler*), in the *Department of Mathematics and Computer Science* of the *University of Haifa*, Israel (particularly *Yaki Sternfeld, Jonathan Arazi, Yair Censor, Jonathan Golan*, and *Izu Vaisman*), and in the *Department of Mathematics* of *The University of Texas at Arlington* (chairman *George Fix*). Their help in providing both the scientific environment for our cooperation and the financial support for our mutual visits to Linz and Haifa, as well as for the visit of the first author to Arlington, has made this book possible.

We very much appreciate the professional support of *Waltraud Eidljörg* in typing several versions of the manuscript, and of *Daria Nováková* in providing the final version in LaTeX. *Gerard Wearden* very much helped us when problems with the English language occured. All Figures have been drawn using MATHEMATICA™.

Special thanks go to the *Editor* of this Series, *Stephen Tijs*, and to *Maria M. Stratta* from *Kluwer Academic Publishers*.

Haifa and Linz, December 1992

Dan Butnariu
Erich Peter Klement

Introduction

This book aims to present, in a unified approach, a series of mathematical results concerning *triangular norm–based measures* and a class of cooperative *games with fuzzy coalitions.* Our approach intends to emphasize that triangular norm–based measures are powerful tools in exploring the coalitional behaviour in such games. They not only unify and simplify some technical aspects of the already classical axiomatic theory of *Aumann–Shapley values*, but also provide new perspectives and insights into these results. Moreover, this machinery allows us to obtain, in the game theoretical context, new and heuristically meaningful information, which has a significant impact on *balancedness* and *equilibria* analysis in a cooperative environment.

From a formal point of view, *triangular norm-based measures* are valuations on subsets of a unit cube $[0,1]^X$ which preserve dual binary operations induced by *triangular norms* on the unit interval $[0,1]$. Triangular norms (and their dual conorms) are algebraic operations on $[0,1]$ which were suggested by MENGER [1942] and which proved to be useful in the theory of *probabilistic metric spaces* (see also [WALD 1943]). The idea of a triangular norm–based measure was implicitly used under various names: *vector integrals* [DVORETZKY, WALD & WOLFOWITZ 1951], *probabilities of fuzzy events* [ZADEH 1968], and *measures on ideal sets* [AUMANN & SHAPLEY 1974, p. 152].

Games with fuzzy coalitions are described in this book by special functions on subsets of a unit cube $[0,1]^X$. Particular classes of such games considered previously were called *ideal set functions* [AUMANN & SHAPLEY 1974, p. 144], *fuzzy games* [AUBIN 1974 a,b; 1982, p. 320], *generalized games* [AUBIN 1981] or, simply, *games* [RUCKLE 1982, 1987]. They were studied because of their relevance to the interaction between values, cores and equilibria, an interaction which became self-evident in [AUMANN & SHAPLEY 1974, Chapters IV — VI]. The connection between cooperative games and certain classes of triangular norm–based measures was implicitly noted by DVORETZKY, WALD & WOLFOWITZ [1950], SHAPLEY [1967] and AUMANN & SHAPLEY [1974, p. 152]. The present book is a first attempt to exploit this relationship.

The proper domain for a game with fuzzy coalitions is a *triangular norm–based tribe* which, in our interpretation, is a collection of *fuzzy coalitions*, the formation and dismantling of which obey some exogeneous rules modeled in terms of algebraic operations induced by triangular norms.

The term *fuzzy coalition*, first introduced in [AUBIN 1974a, b], denotes a collection of economic agents, i.e., *players*, who transfer *fractions of* [their] *representability* [BILLOT 1990] to a collective decision maker, the fuzzy coalition. The fraction of representability a player transfers to a specific coalition is called the *degree of membership* (of the player in that coalition), and it may be any real number in the interval $[0,1]$.

Formally, fuzzy coalitions are modelled by *fuzzy sets* [ZADEH 1965], i.e., functions

$A : X \longrightarrow [0,1]$, where X is the set of players and $A(x)$ represents the *degree of membership* of the player x in the fuzzy coalition A. However, the fuzzy coalitions are closely linked to a fixed underlying triangular norm–based tribe $\mathscr{T}$, i.e., we shall use the term fuzzy coalition only for fuzzy sets which are contained in $\mathscr{T}$. This automatically imposes a measurability condition upon the function $A : X \longrightarrow [0,1]$ (see Theorem 3.2), where fuzzy coalitions are proper generalizations of *ideal sets* in the sense of AUMANN & SHAPLEY [1974, p. 142]. It turns out that tribes of ideal sets are exactly what we shall call *generated tribes.*

The concept of a fuzzy coalition, i.e., a collective decisional entity where members may have gradual degrees of membership, may be somewhat surprising for scientists used to working with the concept of coalition as a collection of players who are indistinguishable from each other except for their marginal contribution to the total *worth* of the coalition, as it emerged from the work of NEUMANN & MORGENSTERN [1944] and most of the literature based on it. Such crisp collections of players will be called *crisp coalitions*, and they are, in fact, particular *fuzzy* coalitions.

Of course, in many real cooperative games only crisp coalitions are possible. However, in other cases the restriction to crisp coalitions only is an excessive idealization. For instance, consider a cooperative game in which each coalition is a group of shareholders of a company. The worth of a coalition is determined by the total investment (endowment) of its members in this company. Assume that two members of a coalition invest the same amount (and, consequently, have the same share in and the same marginal contribution to this coalition) and assume that this amount represents 90 % of the individual worth of the first player and only 5 % of the second player's individual worth. In such a case, it is hard to imagine that the two players will have equal interest in the performance of the coalition because of their different degrees of personal involvement. The use of fuzzy coalitions makes it easy to represent the different involvement of the two players.

Some authors have linked the introduction of fuzzy coalitions to the *convexification* of a set of crisp coalitions. In our general approach, the convexity of a tribe of coalitions plays little role: a tribe is a convex set of functions if and only if it is generated, but the consistency of the solution concepts we deal with does not depend on whether the tribe is generated or not (see Theorem 18.4).

From a game theoretical point of view, the main concept investigated in this work is that of *Aumann–Shapley value.* We use it in the more specific framework of games with fuzzy coalitions, at the same time preserving the original *plausibility conditions: efficiency, symmetry, positivity* and *additivity.* Aumann–Shapley values for games with fuzzy coalitions have the same natural interpretation as the original concept [AUMANN & SHAPLEY 1974, p. 15].

The principal question is whether *diagonal* Aumann–Shapley values for games with fuzzy coalitions exist and, if so, how they can be represented in an analytical way. We start our discussion at the point which is marked by the contributions of AUMANN & SHAPLEY [1974], MERTENS [1980, 1988] and RUCKLE [1982, 1987], and we show that the domain of diagonal Aumann–Shapley values can be expanded further.

In order to do so, we heavily exploit the very special nature of measures based on *fundamental triangular norms*, since these measures allow disintegrations via *Markov kernels.* This disintegration is possible because of the built–in measurability structure of tribes with respect to such triangular norms (see Theorem 3.2), and it is proved for generated tribes in Theorems 5.8 and 7.1 and for nongenerated tribes in Theorems 6.2 and 8.6. It may be that equally powerful representations can be obtained for other classes of triangular norm–based measures, in which case our results could be expanded even further. However, a crucial point in our game theoretical considerations is the validity of a Liapounoff type theorem (see Theorem 14.2) for a large class of triangular norm–based measures.

Using these facts we establish the existence of Aumann–Shapley values which are represented as integrals of various differential operators on the diagonal of a unit cube $[0,1]^X$ (diagonal values) without imposing the *standardness assumption.* Such values are defined on different spaces of games which are included in or overlapping with each other, as illustrated in the figure below. In this figure, *pNA* denotes the usual closed space of games of bounded variations spanned by linear combinations of powers of nonatomic probability measures on a given σ-algebra of crisp coalitions [Aumann & Shapley 1974, p. 22], which here is identified with its isomorphic image through the *extension operator* (see Section 19). The space *pFNA* is the analogously constructed space of games with fuzzy coalitions [Butnariu 1987], *GDIFFE* and *GDIFFE** are the spaces of games which are Gâteaux and Clarke differentiable almost everywhere along the diagonal of $[0,1]^X$, respectively (see Sections 23 and 24). The spaces *MDIFFE* and *MDIFFE** are the domains of the Aumann–Shapley value operators for games with fuzzy coalitions, as implicitly defined by Mertens [1980, 1988]. The intersection of *FBV* with Ruckle's [1987] space of games with fuzzy coalitions C_0^1 is a proper subspace of *GDIFFE*, and the value operator we define on *GDIFFE* is a continuous extension of Ruckle's value on C_0^1.

Thinking of applications, the theoretical results in this book can be used in various ways. On the one hand, they provide methods to enlarge the domain of Aumann–Shapley values for games with crisp coalitions (Section 26). Another application could be the balancedness analysis in the spirit of Aumann & Perles [1965] (see Section 27). Aumann–Shapley values are also useful in solving real-life problems such as rate problems for services in bulk [Billera, Heath & Raanan 1978] as discussed in Section 28. In this work we mostly treat triangular norm–based measures as tools in game theory. However, they can be differently interpreted in different contexts. For instance, it is possible to think of them in logical terms (see Section 29).

In the course of preparing the manuscript some open problems occured which, together with bibliographical comments, are summarized in Section 30 and which indicate possible directions for further investigations.

Our presentation of the two main topics, triangular norm–based measures and games with fuzzy coalitions, is by no means exhaustive. We mostly restrict ourselves to the interrelation between the representation of a class of measures and the existence of Aumann–Shapley values for games in characteristic function form with some smoothness properties along the diagonal. Therefore, important contributions to both areas are not included here. For instance, we do not deal with the measures introduced by SUGENO [1974] and HÖHLE [1976] which are also defined on subsets of the unit cube, nor with the decomposable measures introduced by WEBER [1985a, b], where triangular conorms are used in a different way. On the other hand, the basic works of AUBIN [1974a, b; 1981] about games with fuzzy coalitions are not discussed here since they are already summarized authoritatively by AUBIN [1982] himself. Other important results in game theory (e.g., [HART 1980], [MONDERER & NEYMANN 1987], [TAUMANN 1981]) were omitted in spite of the fact that they can easily be reinterpreted within our framework since we are convinced that such a reinterpretation per se does not provide significant new information. Our treatment of games with fuzzy coalitions provides a method of dealing with vagueness in a game theoretical context. There are other ways of incorporating vague information in such models (see [FUSTIER 1986], [PONSARD 1986a, b], among others).

Finally, our original motivation to write this book was to look from a different perspective at fundamental concepts in measure and game theory, a perspective which we feel is revealing and productive. Our personal interest in this subject was ignited by the mathematical elegance and economic significance of the theory developed in the book by AUMANN & SHAPLEY [1974] as well as by the suggestive and philosophical power of the concept of fuzzy sets as described by ZADEH [1965].

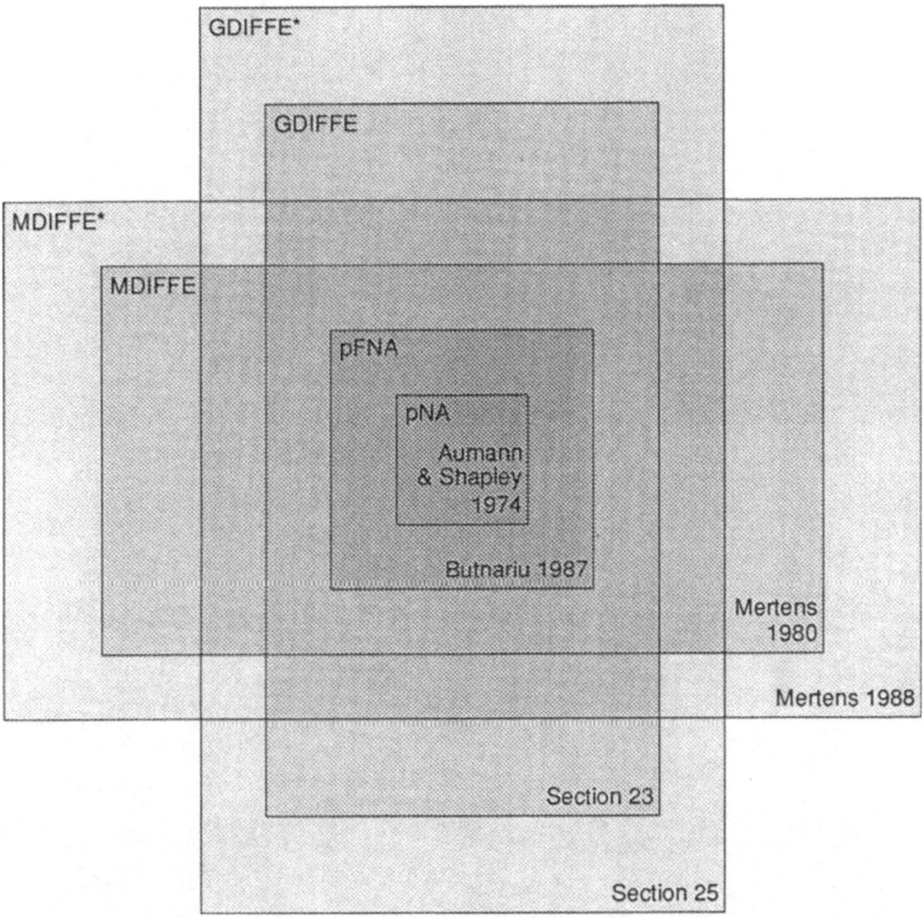

Different spaces with Aumann–Shapley values

Chapter I

Triangular Norm–Based Tribes

1 Triangular Norms

1.1 The class of cooperative games, which is the main topic of this book and which we shall investigate in detail in chapters IV — VI, consists of games with *fuzzy coalitions*. In such coalitions players are allowed to belong in a *gradual way* (as opposed to the *yes/no alternative* in classical models of games in characteristic function form, as described by NEUMANN & MORGENSTERN [1944]). Fuzzy coalitions will be described by *fuzzy sets* [ZADEH 1965], a generalization of *Cantorian sets*. For a Cantorian set A the possible truth values of the statement "x is an element of A" are 0 and 1 (*False* and *True*), whereas for a fuzzy set A the truth values of the statement "x is an element of A" can be any number in the interval $[0, 1]$. We shall describe fuzzy sets in more detail in the next section; here we want to provide the tools for extending operations such as intersection and union of sets. This enables us to represent algebraically natural entropic rules for forming and dismantling fuzzy coalitions.

In order to do so properly, we consider suitable functions $T : [0,1] \times [0,1] \longrightarrow [0,1]$ and $S : [0,1] \times [0,1] \longrightarrow [0,1]$, whose restrictions to $\{0,1\} \times \{0,1\}$ coincide with conjunction and disjunction, respectively, and which satisfy such general properties like associativity, commutativity and monotonicity.

Such functions are known as triangular norms and triangular conorms, respectively; they proved to be useful in the theory of *probabilistic metric spaces*([MENGER 1942], [SCHWEIZER & SKLAR 1983]). In such spaces the distance between two points is measured by probability distribution functions rather than by real numbers, and the triangular norms allow the proper generalization of the triangle inequality.

In *continuum physics* MENGER [1951a, b] interpreted triangular norms as *rules* for generating new probabilistically determined objects from existing ones in the psychophysical continuum space. In accordance with one of the main ideas underlying the foundation of modern game theory by NEUMANN & MORGENSTERN [1944], we shall use triangular norms as models for interaction rules of economic agents in a way which is similar to the modelling of the interaction of physical particles.

Triangular norms are also closely related to the still unsolved general associativity functional equation [CLIMESCU 1946].

1.2 Definition.

(i) A function $T : [0,1] \times [0,1] \longrightarrow [0,1]$ is called *triangular norm* (*t-norm* for short) if it satisfies the following conditions:

Boundary condition

$$T(x,1) = x, \tag{1.1}$$

Monotonicity

$$T(x,y) \leq T(x,z) \quad \text{whenever} \quad y \leq z, \tag{1.2}$$

Commutativity

$$T(x,y) = T(y,x), \tag{1.3}$$

Associativity

$$T(x,T(y,z)) = T(T(x,y),z). \tag{1.4}$$

(ii) A t-norm T is called *Archimedean* if it is continuous and satisfies

$$T(x,x) < x \quad \text{whenever} \quad 0 < x < 1. \tag{1.5}$$

(iii) A t-norm T is called *strict* if it is continuous and satisfies

$$T(x,y) < T(x,z) \quad \text{whenever} \quad x > 0 \text{ and } y < z. \tag{1.6}$$

(iv) If T is a triangular norm, then the *corresponding triangular conorm* (*t-conorm*) is the function $S : [0,1] \times [0,1] \longrightarrow [0,1]$ defined by

$$S(x,y) = 1 - T(1-x,1-y). \tag{1.7}$$

1.3 Remark.

(i) From the definition the following additional properties of a t-norm T are derived immediately:

$$\begin{aligned} T(x,0) &= 0, & (1.8)\\ T(x,y) &\leq T(u,v) \quad \text{whenever} \quad x \leq u \quad \text{and} \quad y \leq v. & (1.9) \end{aligned}$$

(ii) If T is a strict t-norm then

$$T(x,y) < T(u,v) \quad \text{whenever} \quad x < u \quad \text{and} \quad y < v. \tag{1.10}$$

(iii) Each strict t-norm is Archimedean; the converse is not true in general (e.g., see the t-norm T_∞ in Example 1.4(i)).

(iv) The corresponding t-conorm S of a t-norm T satisfies (1.2), (1.3), (1.4) and, dually to (1.1),

Boundary condition

$$S(x,0) = x. \tag{1.11}$$

(v) A function $S : [0,1] \times [0,1] \longrightarrow [0,1]$ fulfilling (1.11), (1.2), (1.3) and (1.4) is called a *triangular conorm* (*t-conorm*).

(vi) It is easily seen that given a t-conorm S, then $T : [0,1] \times [0,1] \longrightarrow [0,1]$ defined by

$$T(x,y) = 1 - S(1-x, 1-y). \tag{1.12}$$

is a t-norm whose corresponding t-conorm is exactly the t-conorm S we started with.

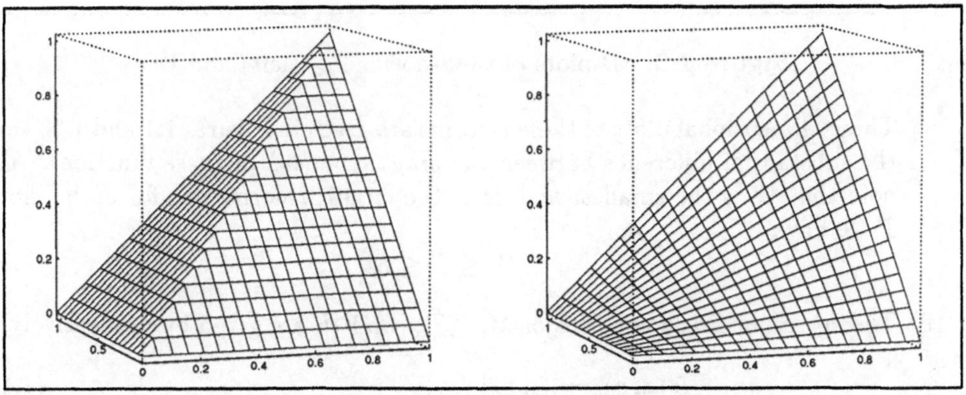

Figure 1.1: 3D plots of the t-norms M (left) and P

1.4 Example.

(i) Some of the most important t-norms are the functions M, P, T_∞ and W defined as follows:

$$\begin{aligned} M(x,y) &= \min(x,y), && (1.13)\\ P(x,y) &= x.y, && (1.14)\\ T_\infty(x,y) &= \max(0,x+y-1), && (1.15)\\ W(x,y) &= \begin{cases} \min(x,y) & \text{if } \max(x,y)=1,\\ 0 & \text{otherwise.}\end{cases} && (1.16)\end{aligned}$$

The t-norm M is not Archimedean (and hence not strict), P is strict (and therefore Archimedean), T_∞ is Archimedean but not strict, and W, not being continuous, is neither Archimedean nor strict.

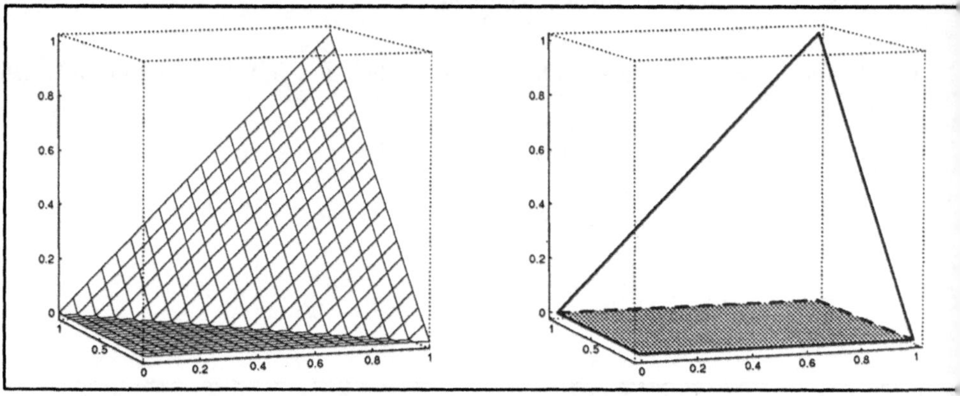

Figure 1.2: 3D plots of the t-norms T_∞ (left) and W

Three–dimensional plots of these t-norms are given in Figures 1.1 and 1.2. Note the substantial difference between the *graphic outlook* of these functions. Also note that W is the smallest and M is the largest t-norm, i.e., for each t-norm T we have

$$W \leq T \leq M. \qquad (1.17)$$

(ii) The corresponding t-conorms of M, P, T_∞ and W are given by, respectively,

$$\begin{aligned} N(x,y) &= \max(x,y), && (1.18)\\ O(x,y) &= x+y-x.y, && (1.19)\\ S_\infty(x,y) &= \min(1,x+y), && (1.20)\end{aligned}$$

$$V(x,y) = \begin{cases} \max(x,y) & \text{if } \min(x,y)=0, \\ 1 & \text{otherwise.} \end{cases} \tag{1.21}$$

1.5 A very special role is played in this book by the following t-norms which we therefore call fundamental t-norms:

Definition.

(i) For each $s \in [0,\infty]$ the *fundamental t-norm* T_s is given by

$$T_s(x,y) = \begin{cases} M(x,y) & \text{if } s=0, \\ P(x,y) & \text{if } s=1, \\ T_\infty(x,y) & \text{if } s=\infty, \\ \log_s\left[1+\dfrac{(s^x-1)\cdot(s^y-1)}{s-1}\right] & \text{if } s\in\,]0,\infty[\,\setminus\{1\}. \end{cases} \tag{1.22}$$

(ii) The *fundamental t-conorms* $(S_s)_{s\in[0,\infty]}$ are defined as follows:

$$S_s(x,y) = \begin{cases} N(x,y) & \text{if } s=0, \\ O(x,y) & \text{if } s=1, \\ S_\infty(x,y) & \text{if } s=\infty, \\ 1-\log_s\left[1+\dfrac{(s^{1-x}-1)\cdot(s^{1-y}-1)}{s-1}\right] & \text{if } s\in\,]0,\infty[\,\setminus\{1\}. \end{cases} \tag{1.23}$$

A fundamental t-norm T_s is strict if and only if $0 < s < \infty$, T_∞ is Archimedean but not strict, and T_0 is neither Archimedean nor strict.

The family of t-norms $(T_s)_{s\in[0,\infty]}$ and their corresponding t-conorms $(S_s)_{s\in[0,\infty]}$ has been studied by FRANK [1979]. It is a *continuous* family in the sense that for each $t \in [0,\infty]$ we have

$$\lim_{s\to t} T_s = T_t\,. \tag{1.24}$$

The *diagonal plots* shown in Figure 1.7 provide graphs of the diagonal functions $x \longrightarrow T_s(x,x)$ for some fundamental t-norms.

Moreover, each pair (T_s, S_s) of mutually corresponding fundamental t-norms and t-conorms satisfies the functional equation

$$T(x,y)+S(x,y) = x+y. \tag{1.25}$$

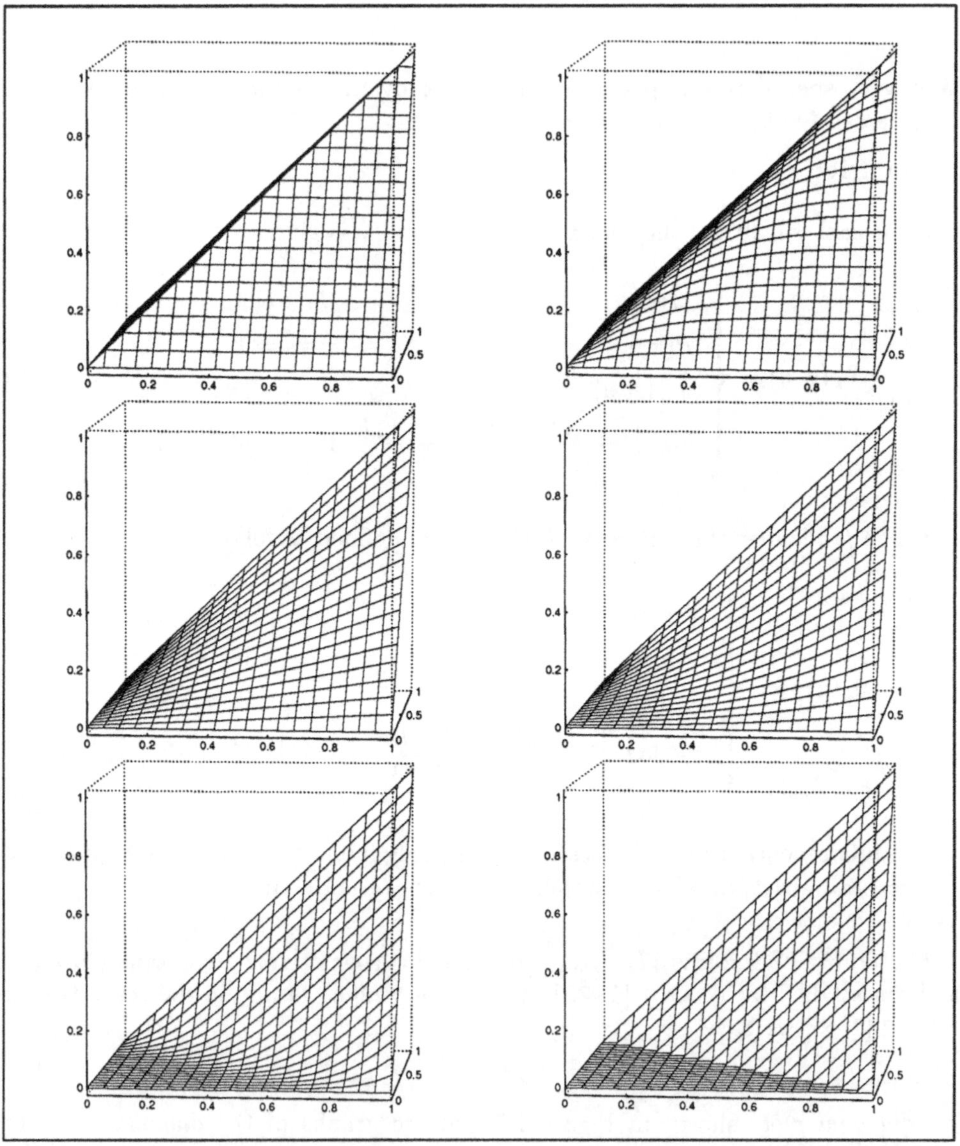

Figure 1.3: 3D plots of some fundamental t-norms: T_0 (top left), $T_{0.001}$ (top right), T_1 (center left), T_{10} (center right), $T_{1000,000}$ (bottom left), T_∞ (bottom right)

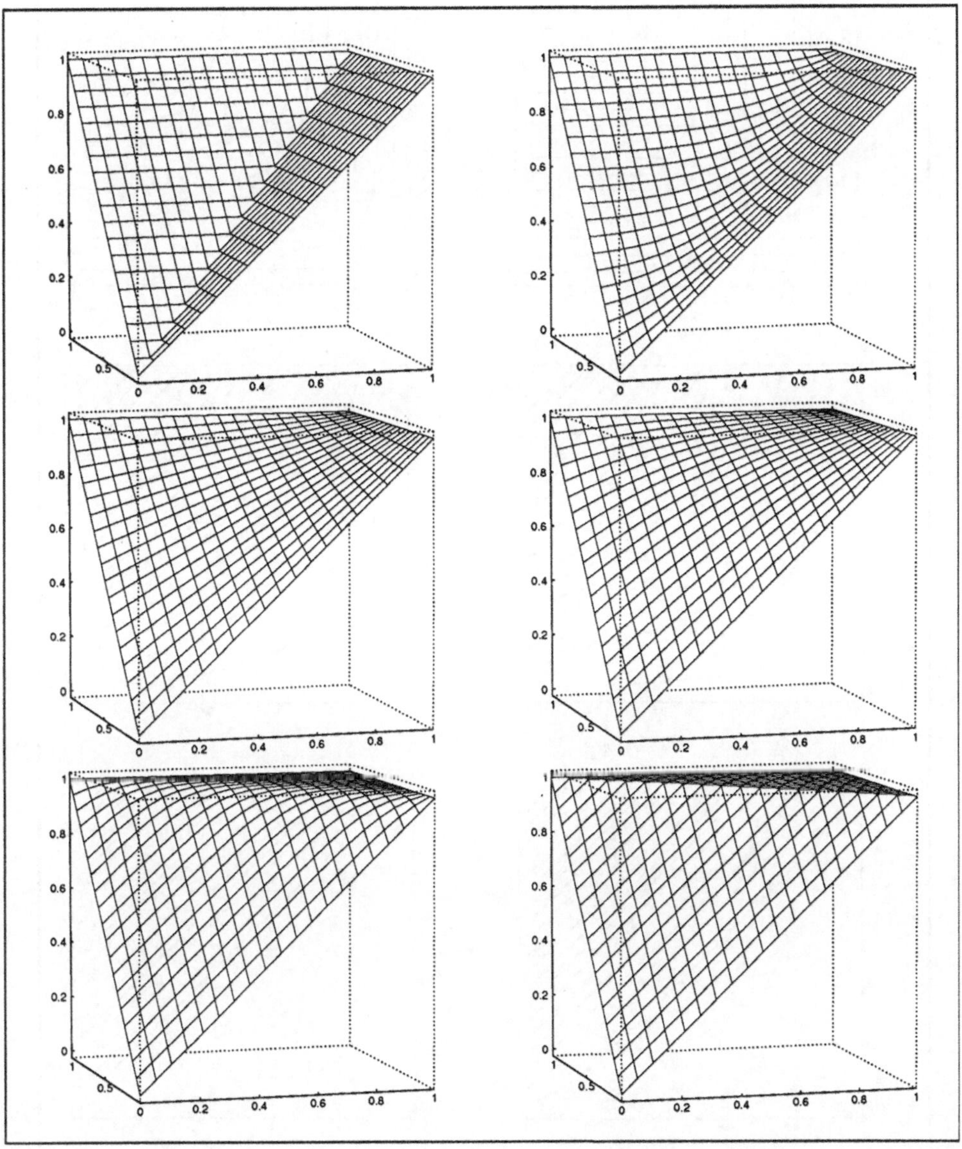

Figure 1.4: 3D plots of some fundamental t-conorms: S_0 (top left), $S_{0.001}$ (top right), S_1 (center left), S_{10} (center right), $S_{1000,000}$ (bottom left), S_∞ (bottom right)

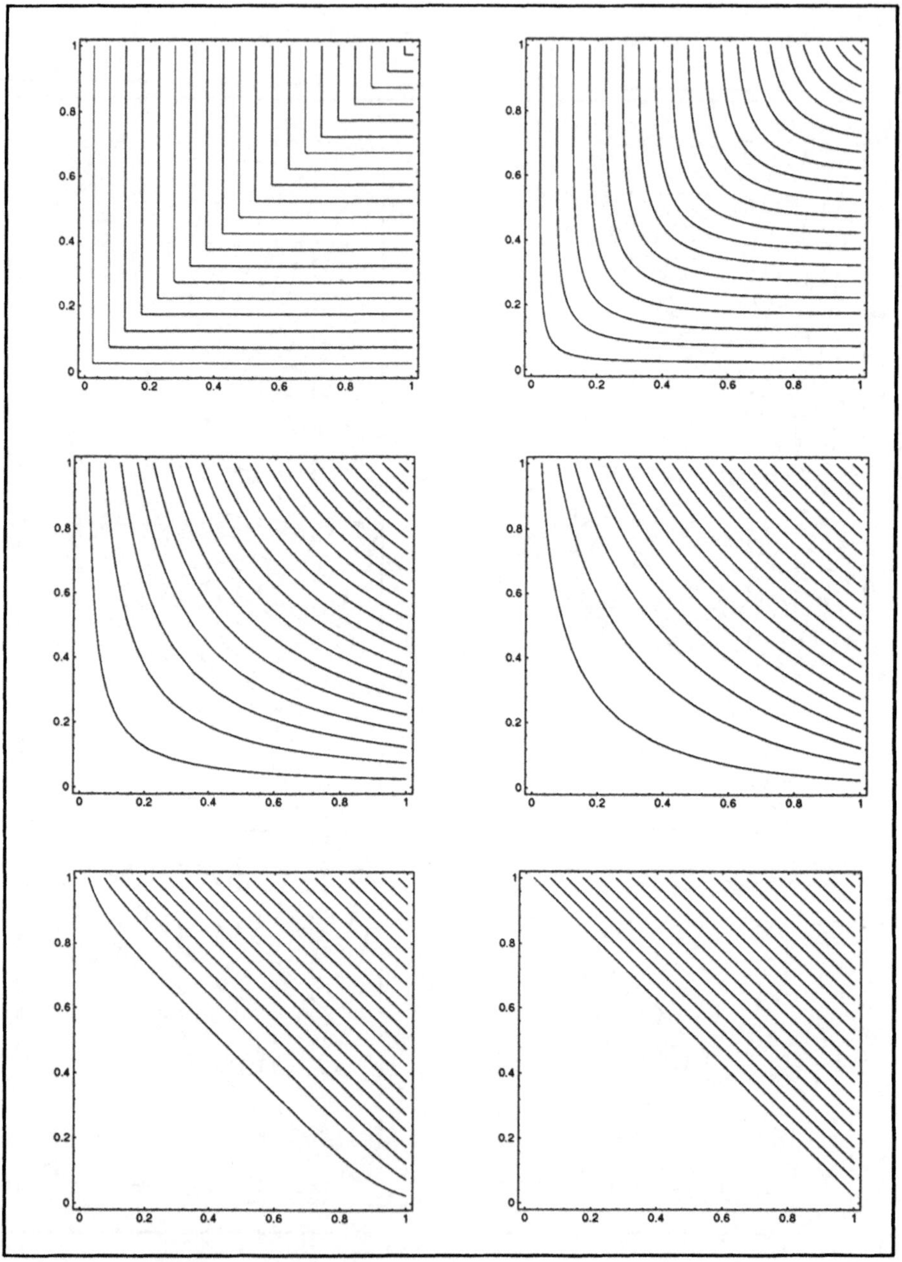

Figure 1.5: Contour plots of some fundamental t-norms: T_0 (top left), $T_{0.001}$ (top right), T_1 (center left), T_{10} (center right), $T_{1000,000}$ (bottom left), T_∞ (bottom right)

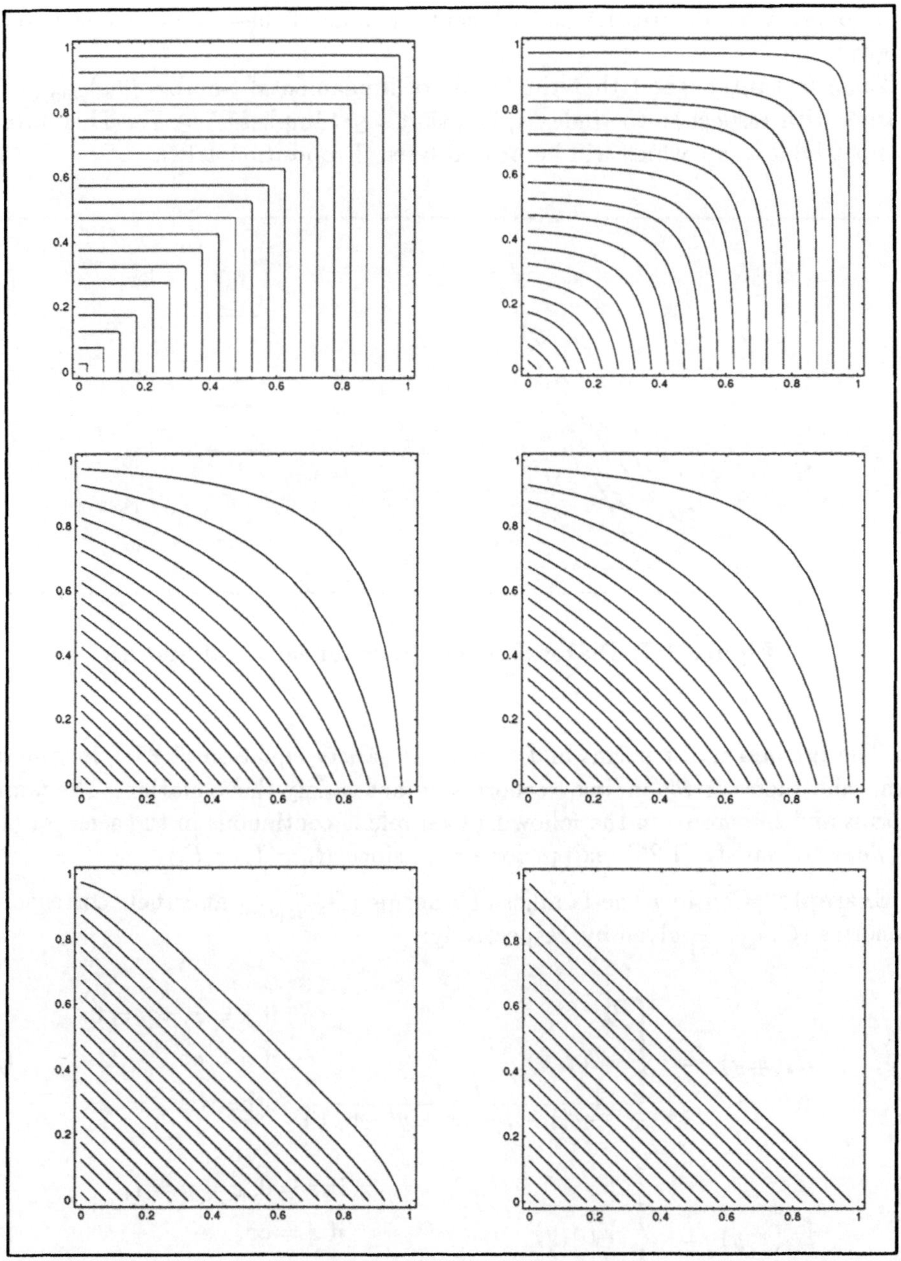

Figure 1.6: Contour plots of some fundamental t-conorms: S_0 (top left), $S_{0.001}$ (top right), S_1 (center left), S_{10} (center right), $S_{1000,000}$ (bottom left), S_∞ (bottom right)

Figures 1.3 and 1.4 provide three-dimensional plots of some fundamental t-norms and t-conorms. In Figures 1.5 and 1.6 contour plots of these t-norms and t-conorms are given.

These pictures suggest that the family of fundamental t-norms $(T_s)_{s\in[0,\infty]}$ is decreasing with respect to the index s, i.e., that $s \le t$ implies $T_s \ge T_t$. This, however, is a nontrivial result which will be proved later (Proposition 1.12).

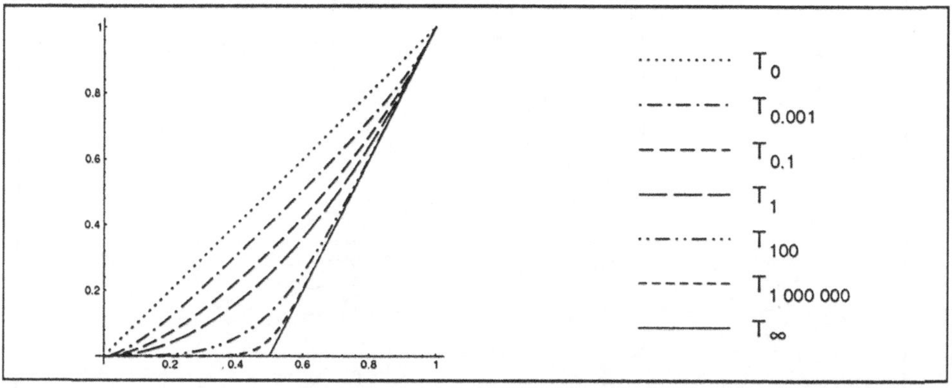

Figure 1.7: Diagonal cuts of some fundamental t-norms

1.6 The fundamental t-norms and t-conorms satisfy equation (1.25). In general, t-norms and their corresponding t-conorms do not satisfy this equation. The family of t-norms and t-conorms in the following example is continuous in the sense of (1.24), but does not satisfy (1.25), except for $s = 1$, since $H_1 = T_1 = P$.

Example. Consider the families of t-norms $(H_s)_{s\in[0,\infty]}$ and their corresponding t-conorms $(G_s)_{s\in[0,\infty]}$ given by, respectively:

$$H_s(x,y) = \begin{cases} 0 & \text{if } s = x = y = 0, \\ W(x,y) & \text{if } s = \infty, \\ \dfrac{x \cdot y}{s + (1-s)\cdot(x + y - x\cdot y)} & \text{otherwise,} \end{cases} \tag{1.26}$$

$$G_s(x,y) = \begin{cases} 1 & \text{if } s = 0, x = y = 1, \\ V(x,y) & \text{if } s = \infty, \\ \dfrac{x + y + (s-2)\cdot x\cdot y}{1 + (s-1)\cdot x \cdot y} & \text{otherwise.} \end{cases} \tag{1.27}$$

The family of t-norms $(H_s)_{s\in[0,\infty]}$ has been studied by HAMACHER [1978] in an attempt to provide an axiomatic basis for operations on fuzzy sets. It is a decreasing family, i.e., for $0 < s < t < \infty$ we have $H_0 > H_s > H_t > H_\infty$, and it is also continuous

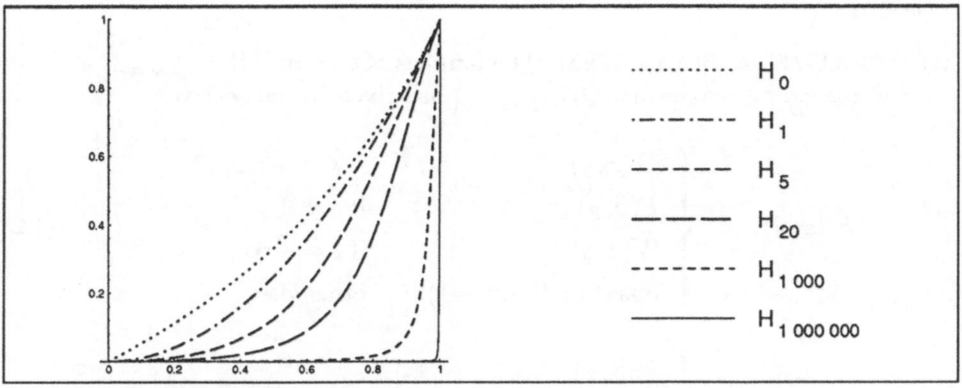

Figure 1.8: Diagonal cuts of some t-norms in $(H_s)_{s\in[0,\infty]}$

in the sense of (1.24). With the exception of H_∞ (which is not even Archimedean), all the t-norms H_s are strict.

Diagonal cuts and three-dimensional plots of several t-norms H_s are given in Figures 1.8 and 1.10, respectively. Note the strong similarity between the geometric forms of the t-norms H_s and T_s.

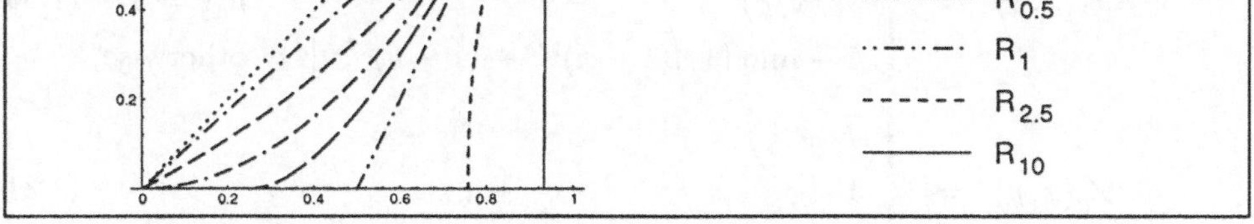

Figure 1.9: Diagonal cuts of some t-norms $(R_s)_{s\in[-\infty,\infty]}$

1.7 Interesting continuous (in the sense of (1.24)) and decreasing families of t-norms, which look quite dissimilar from the fundamental t-norms, are given in the following (for more details concerning these and other classes of t-norms, see [KLIR & FOLGER 1988]).

Example.

(i) [SCHWEIZER & SKLAR 1983]. The families of t-norms $(R_s)_{s\in[-\infty,+\infty]}$ and their corresponding t-conorms $(Q_s)_{s\in[-\infty,+\infty]}$ are given by, respectively:

$$R_s(x,y) = \begin{cases} M(x,y) & \text{if } s=-\infty, \\ P(x,y) & \text{if } s=0, \\ W(x,y) & \text{if } s=+\infty, \\ [\max(0,x^s+y^s-1)]^{1/s} & \text{otherwise,} \end{cases} \tag{1.28}$$

$$Q_s(x,y) = \begin{cases} N(x,y) & \text{if } s=-\infty, \\ O(x,y) & \text{if } s=0, \\ V(x,y) & \text{if } s=+\infty, \\ 1-[\max(0,(1-x)^s+(1-y)^s-1)]^{1/s} & \text{otherwise.} \end{cases} \tag{1.29}$$

The *diagonal cuts* and the 3D plots of several t-norms R_s are given in Figure 1.9 and Figure 1.11, respectively. Note that the t-norms M, P, T_∞ and W are either a member or a limit of the family $(R_s)_{s\in[-\infty,+\infty]}$.

A t-norm R_s is strict if and only if $-\infty < s \leq 0$, it is non-strict Archimedean if and only if $0 < s < +\infty$; $R_{-\infty}$ and $R_{+\infty}$ are not Archimedean.

(ii) [YAGER 1980]. The family of t-norms $(Y_s)_{s\in[0,\infty]}$ and the corresponding family of t-conorms $(Z_s)_{s\in[0,\infty]}$ are given by, respectively:

$$Y_s(x,y) = \begin{cases} M(x,y) & \text{if } s=0, \\ W(x,y) & \text{if } s=\infty, \\ 1-\min\left(1,\left[(1-x)^{1/s}+(1-y)^{1/s})\right]^s\right) & \text{otherwise,} \end{cases} \tag{1.30}$$

$$Z_s(x,y) = \begin{cases} N(x,y) & \text{if } s=0, \\ V(x,y) & \text{if } s=\infty, \\ \min\left(1,\left[x^{1/s}+y^{1/s})\right]^s\right) & \text{otherwise.} \end{cases} \tag{1.31}$$

Except for Y_0 and Y_∞ (which are not Archimedean) all the t-norms Y_s are Archimedean but not strict. *Diagonal cuts* and 3D plots of several t-norms Y_s are given in Figure 1.12 and Figure 1.13, respectively.

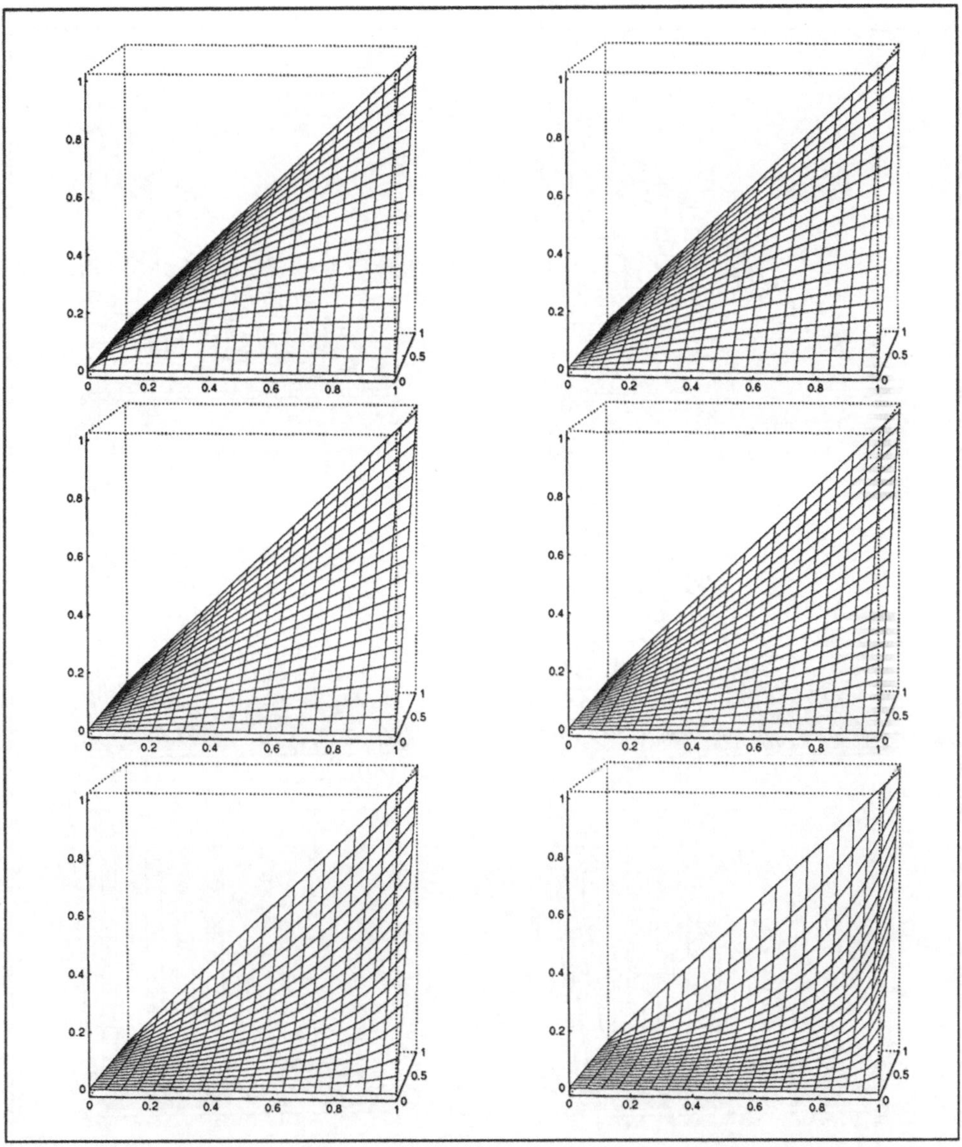

Figure 1.10: 3D plots of some t-norms in $(H_s)_{s\in[0,\infty]}$: H_0 (top left), $H_{0.5}$ (top right), H_1 (center left), H_2 (center right), H_5 (bottom left), H_{15} (bottom right)

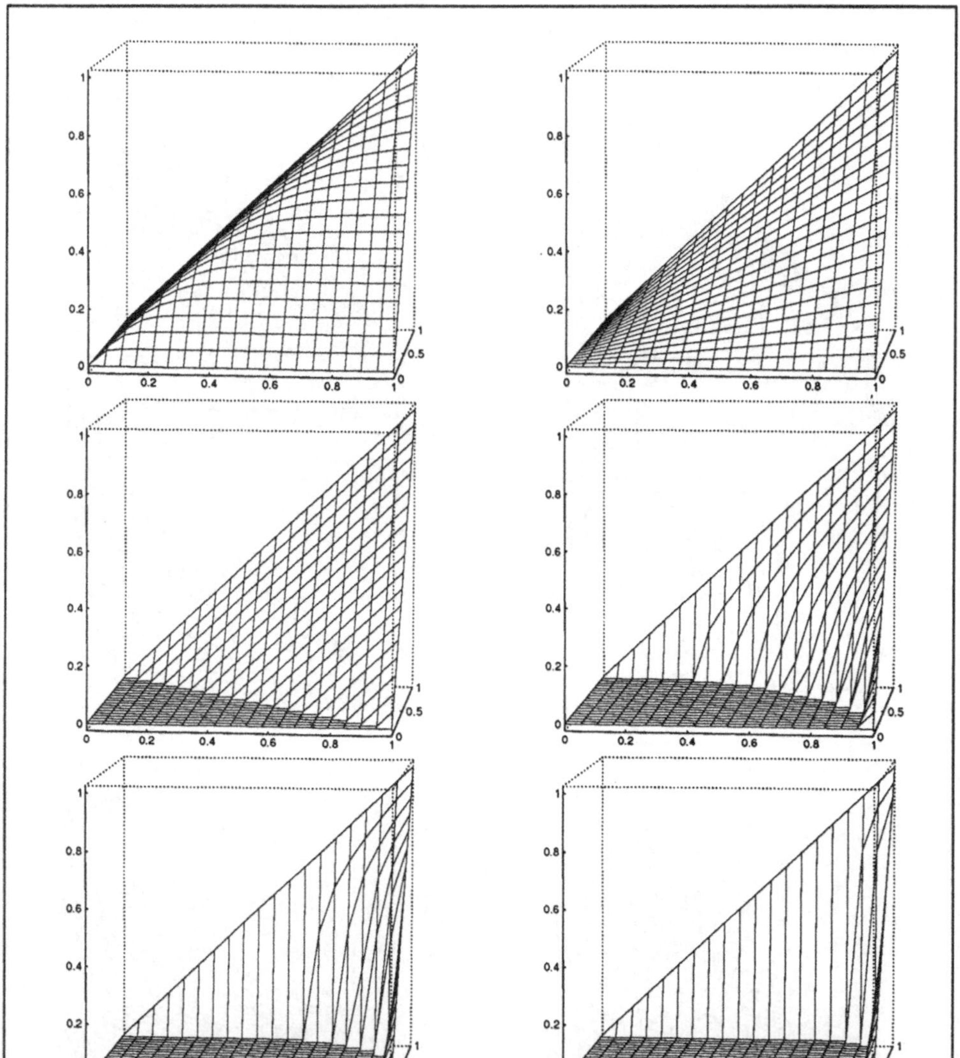

Figure 1.11: 3D plots of some t-norms $(R_s)_{s\in[-\infty,+\infty]}$: R_{-5} (top left), R_0 (top right), R_1 (center left), R_2 (center right), R_4 (bottom left), R_8 (bottom right)

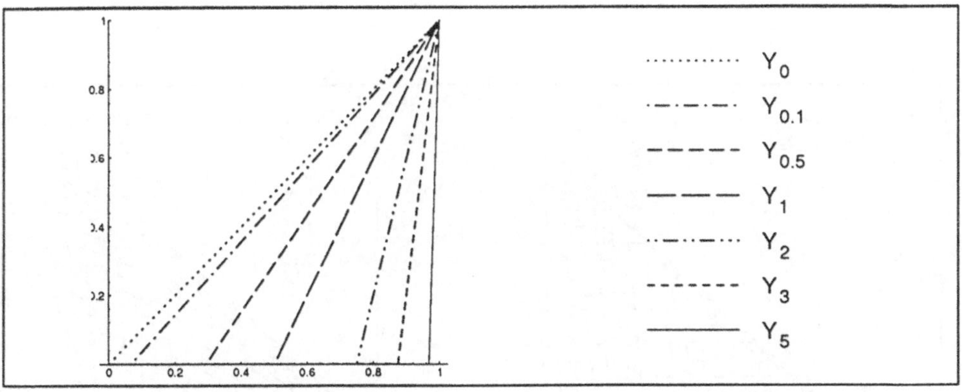

Figure 1.12: Diagonal cuts of some t-norms $(Y_s)_{s\in[0,\infty]}$

1.8 In what follows we mostly deal with t-norms which are either continuous or, at least, Borel measurable functions from $[0,1]\times[0,1] \longrightarrow [0,1]$. In general, a t-norm is not necessarily continuous (e.g., W, as defined in (1.16)); here is an example of a t-norm which is not even Borel measurable:

Example. ([KLEMENT 1982a]). Let A be a non-Borel subset of the interval $]0.75, 1[$ and define

$$B = \{(x,y) \in [0,1]\times[0,1] \mid (x \in A \text{ or } y \in A) \text{ and } (x+y=1.5)\}.$$

The function $T : [0,1]\times[0,1] \longrightarrow [0,1]$ defined by

$$T(x,y) = \begin{cases} 0 & \text{if } (x,y) \in B \text{ or } (x,y \in [0,1[\text{ and } x+y<1.5), \\ 0.5 & \text{if } [(x,y) \notin B \text{ and } x+y=1.5] \\ & \quad \text{or } [x,y \in [0,1[\text{ and } x+y>1.5], \\ \min(x,y) & \text{otherwise,} \end{cases}$$

is a t-norm. However, it is not Borel measurable since the set B is a non-measurable subset of $[0,1]\times[0,1]$.

1.9 The most important results in this book are proven assuming that the t-norms involved are fundamental ones. The fact that the fundamental t-norms T_s are not only continuous, but (together with their corresponding t-conorms S_s) also satisfy equation (1.25), is a substantial technical and structural advantage. FRANK [1979] has shown that a pair (T,S) consisting of a continuous t-norm and its corresponding t-conorm satisfies (1.25) if and only if T is a fundamental t-norm or T is an *ordinal sum* of fundamental t-norms.

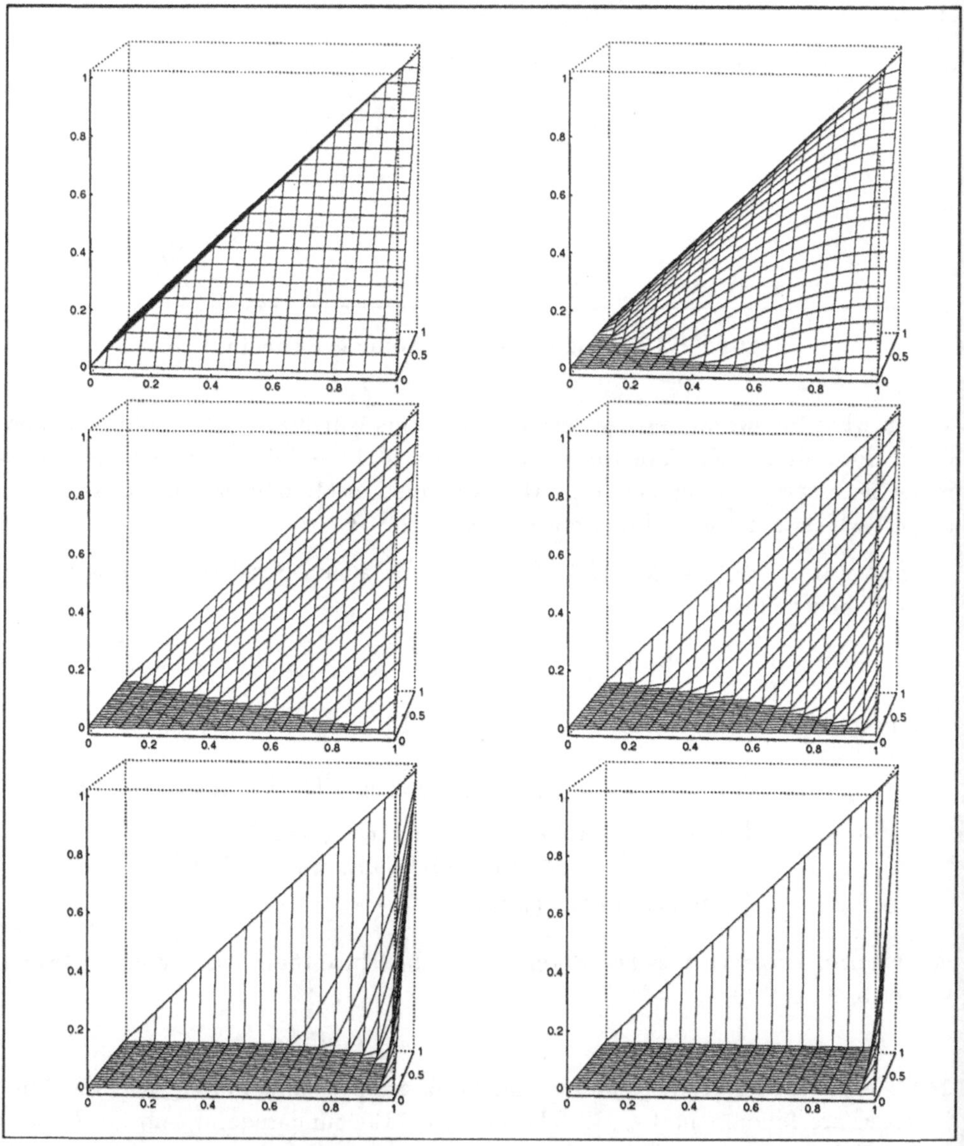

Figure 1.13: 3D plots of some t-norms $(Y_s)_{s\in[0,\infty]}$: Y_0 (top left), $Y_{0.5}$ (top right), Y_1 (center left), $Y_{1.25}$ (center right), $Y_{2.5}$ (bottom left), Y_5 (bottom right)

Given an at most countable family of t-norms $(T_j)_{j\in J}$ and mutually disjoint subintervals $(\,]a_j, b_j[\,)_{j\in J}$ of $[0,1]$, the *ordinal sum* of $(T_j)_{j\in J}$ with respect to $(\,]a_j, b_j[\,)_{j\in J}$ is the function $T : [0,1]\times[0,1] \longrightarrow [0,1]$ defined as follows:

$$T(x,y) = \begin{cases} a_j + (b_j - a_j)\cdot T_j\left(\frac{x-a_j}{b_j-a_j}, \frac{y-a_j}{b_j-a_j}\right) & \text{if } x, y \in\,]a_j, b_j[\text{ for some } j\in J, \\ \min(x,y) & \text{otherwise} \end{cases} \tag{1.32}$$

A three-dimensional plot of an ordinal sum of t-norms is shown in Figure 1.14.

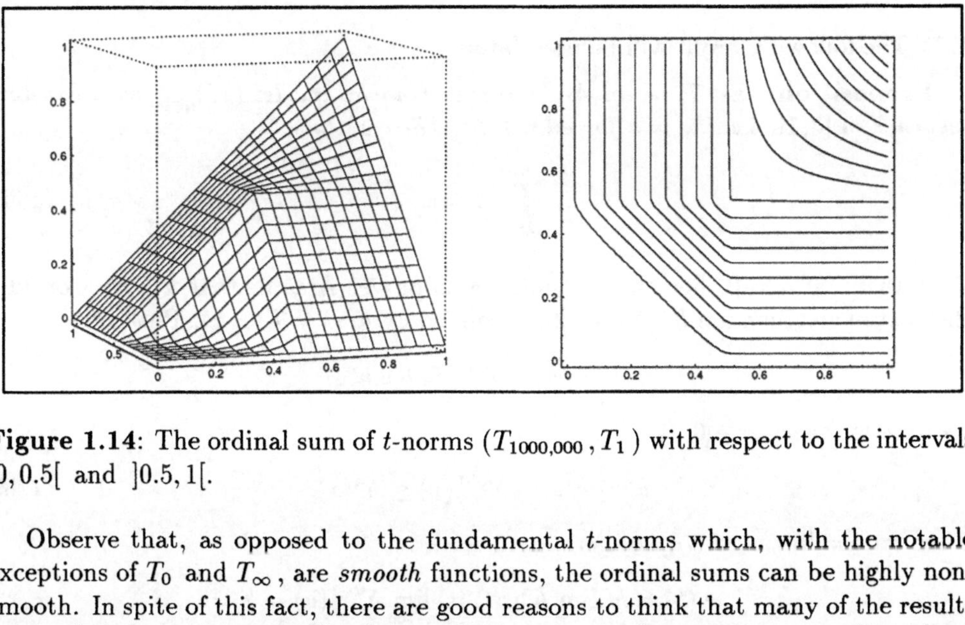

Figure 1.14: The ordinal sum of t-norms $(T_{1000,000}, T_1)$ with respect to the intervals $]0, 0.5[$ and $]0.5, 1[$.

Observe that, as opposed to the fundamental t-norms which, with the notable exceptions of T_0 and T_∞, are *smooth* functions, the ordinal sums can be highly non-smooth. In spite of this fact, there are good reasons to think that many of the results proven below can be generalized by replacing the fundamental t-norms by ordinal sums of fundamental t-norms.

1.10 We may regard a given t-norm T and its corresponding t-conorm S as binary operations on $[0,1]$. Thus, their associativity allows them to be extended to n-ary operations

$$\mathop{\mathrm{T}}_{i=1}^{n} : [0,1]^n \longrightarrow [0,1]$$

and

$$\mathop{\mathrm{S}}_{i=1}^{n} : [0,1]^n \longrightarrow [0,1].$$

In what follows we shall write $T_{i=1}^{n}\, x_i$ and $S_{i=1}^{n}\, x_i$ instead of $T(x_1,\ldots,x_n)$ and $S(x_1,\ldots,x_n)$, respectively. The extension to an infinitary operation is also possible: for each sequence $(x_n)_{n\in\mathbb{N}}$ in $[0,1]$ the sequence $(T_{i=1}^{n} x_i)_{n\in\mathbb{N}}$ is nonincreasing (by (1.17)); therefore its limit

$$\mathop{\mathrm{T}}_{n=1}^{\infty} x_n = \lim_{n\to\infty} \mathop{\mathrm{T}}_{i=1}^{n} x_i \tag{1.33}$$

always exists. By duality, the sequence $(S_{i=1}^{n} x_i)_{i\in\mathbb{N}}$ is nondecreasing and its limit, denoted $S_{n=1}^{\infty} x_n$, exists. Moreover, we have

$$\mathop{\mathrm{S}}_{n=1}^{\infty} x_n = 1 - \mathop{\mathrm{T}}_{n=1}^{\infty} (1 - x_n). \tag{1.34}$$

1.11 The following result will be used later:

Proposition. *Let T be an Archimedean t-norm and let $(x_n)_{n\in\mathbb{N}}$ be a constant sequence in $[0,1[$, i.e., $x_n = a$ for all $n \in \mathbb{N}$. Then we have*

$$\mathop{\mathrm{T}}_{n=1}^{\infty} x_n = 0. \tag{1.35}$$

PROOF: If $a = 0$ then there is nothing to prove. Assume that $a \neq 0$. Consider the continuous function $h : X \longrightarrow [0,1]$ defined by $h(x) = T(x,x)$. Putting

$$h^1 = h, \quad h^{n+1} = h \circ h^n,$$

we have for every $x \in \,]0,1[$

$$h(x) < x, \quad h^{n+1}(x) \leq h^n(x). \tag{1.36}$$

Then for $b = \lim_{n\to\infty} h^n(a)$ we get

$$h(b) = h(\lim_{n\to\infty} h^n(a)) = \lim_{n\to\infty} h^{n+1}(a) = b$$

which implies $b = 0$ because of (1.36). Since the sequence $(h^n(a))_{n\in\mathbb{N}}$ is a subsequence of the convergent (it is bounded and nonincreasing) sequence $(T_{i=1}^{n} x_i)_{n\in\mathbb{N}}$, the result follows. □

1.12 In many of the subsequent considerations the fact that the fundamental t-norms form a decreasing family is essential. This fact was proved by TAKACS [personal communication].

Proposition.

(i) If T is a t-norm which is either fundamental or the ordinal sum of a family of fundamental t-norms, then

$$T_\infty \leq T \leq T_0 \,. \tag{1.37}$$

(ii) For all $s, t \in [0, \infty]$ with $s < t$ we have

$$T_t \leq T_s \,. \tag{1.38}$$

PROOF: In order to show (i), let T be a t-norm which is either fundamental or the ordinal sum of a family of fundamental t-norms. Then $T \leq T_0$ follows from (1.17) and $T_\infty \leq T$ holds since T together with its corresponding t-conorm S satisfies the equation (1.25).

The proof of (ii) will be carried out in three steps. Consider two fundamental t-norms T_s and T_t with $0 \leq s < t \leq \infty$.

Case 1. If $s = 0$ or $t = \infty$ it was shown in (i) that equality (1.38) always holds.

Case 2. Next we show that (1.38) holds when $0 < s < t < 1$ or $1 < s < t < \infty$. In other words, we have to prove that the function $f : D \times [0,1] \times [0,1] \longrightarrow [0,1]$ given by

$$f(s, x, y) = T_s(x, y)$$

is nonincreasing with respect to s in the set $D =]0,1[\cup]1,\infty[$. To this end, observe that for each $s \in D$ we get:

$$\frac{\partial f}{\partial s}(s,x,y) = \frac{1}{(\ln s)^2} \cdot [f_1(s,x,y) - f_2(s,x,y)],$$

where

$$f_1(s,x,y) = \frac{[xs^{x-1}(s^y - 1) + ys^{y-1}(s^x - 1)](s-1) - (s^x - 1)(s^y - 1)}{[1 + \frac{(s^x-1)(s^y-1)}{s-1}](s-1)^2} \cdot \ln s,$$

$$f_2(s,x,y) = \frac{1}{s} \cdot \ln\left[1 + \frac{(s^x - 1)(s^y - 1)}{s - 1}\right].$$

Also, observe that for all $(s,x,y) \in D \times [0,1] \times [0,1]$ we have

$$\frac{\partial f}{\partial s}(s,0,y) = \frac{\partial f}{\partial s}(s,x,0) = \frac{\partial f}{\partial s}(s,1,y) = \frac{\partial f}{\partial s}(s,x,1) = 0.$$

Now fix an arbitrary $s \in]1,\infty[$ and an arbitrary $y \in]0,1[$ and consider the function $h : [0,1] \longrightarrow \mathbb{R}$ defined by

$$h(x) = \frac{\partial f}{\partial s}(s,x,y). \tag{1.39}$$

If we show that the function h has a single relative minimum in $]0,1[$, then because of $h(0) = h(1) = 0$ this implies $h(x) \leq 0$ for all $x \in [0,1]$, which in turn proves $\frac{\partial f}{\partial s}(s,x,y) \leq 0$ for all $(s,x,y) \in D \times [0,1] \times [0,1]$, i.e., $f(s,x,y)$ is nonincreasing with respect to $s \in D$. It is a matter of computation to obtain

$$h'(x) = \frac{s^{x-1}}{((s-1) + (s^x - 1)(s^y - 1))^2}[x(s^y - 1)(s - s^y) + ys^y(s-1) - s(s^y - 1)].$$

Since the expression

$$\frac{s^{x-1}}{((s-1)+(s^x-1)(s^y-1))^2}$$

is always strictly positive, there is only one value of x such that $h'(x)=0$, namely,

$$x_0 = \frac{s(s^y-1) - ys^y(s-1)}{(s^y-1)(s-s^y)}. \tag{1.40}$$

If $s \in \,]0,1[$ and $y \in \,]0,1[$, then obviously $\bar{s} = \frac{1}{s} \in \,]1,\infty[$ and $\bar{y} = 1-y \in \,]0,1[$. Repeating construction (1.39) for $\bar{s}$ and $\bar{y}$, a simple computation shows that this yields the same x_0 in (1.40). This means that we may restrict ourselves to the situation $s \in \,]0,\infty[$.

Now we show that $x_0 \in \,]0,1[$. Observe that if $s > 1$, then for each $y \in \,]0,1[$

$$s^y < (s-1)y + 1. \tag{1.41}$$

Indeed, $g(y) = (s-1)y+1$ is the affine function joining the points $(0,1)$ and $(1,s)$ of the graph of the convex function s^y. Since the denominator in (1.40) is strictly positive, the inequality $x_0 > 0$ is equivalent to

$$s(s^y-1) - ys^y(s-1) > 0$$

which, in turn, is equivalent to

$$s^{1-y} < s + (1-s)y,$$

the latter being equivalent to (1.41). In order to show that $x_0 < 1$, note that this inequality is equivalent to

$$s(s^y-1) - ys^y(s-1) < (s^y-1)(s-s^y)$$

since the denominator in (1.40) is positive. This last inequality is again equivalent to (1.41), which has already been proven, and, therefore, it is true for each $s > 1$. Evaluation of the second derivative of h at x_0 yields

$$h''(x_0) = \frac{s^{x_0-1}(s^y-1)(s-s^y)}{((s-1)+(s^{x_0}-1)(s^y-1))^2},$$

which turns out to be strictly positive, showing that the only relative extremum of h in x_0 is indeed a minimum, thus completing the proof in this case.

Case 3. It remains to show that for $s < 1 < t$ we have $T_s \geq T_1 \geq T_s$. This is an immediate consequence of Case 2 and the continuity (1.24) since

$$\lim_{r \nearrow 1} T_r = \lim_{r \searrow 1} T_r = T_1 .$$

□

2 Triangular Norm–Based Tribes

2.1 The concept of *triangular norm–based tribes* (*on a set* X), which will play a significant role in this work, is a natural generalization of the notion of σ-algebra of subsets (of X). A triangular norm–based tribe is a very special collection of *fuzzy subsets of* X. By a *fuzzy subset* of X or, simply, a *fuzzy set*, we mean a function $A : X \longrightarrow [0,1]$ where $A(x)$ is interpreted as the *degree of membership* of x to A ([ZADEH 1965]). We identify the subsets of X with their characteristic functions. Therefore, if M is a subset of X, then the formulas "$x \in M$" and "$M(x) = 1$" are interchangeable. If we want to emphasize that a fuzzy set is a characteristic function of a subset of X, we shall call it a *crisp set.*

In this section we study tribes which are based on fundamental t-norms. We show that fundamental triangular norm-based tribes are necessarily T_∞-tribes and T_0-tribes. An immediate consequence of this fact is that fundamental triangular norm-based tribes are also *clans* in the sense of WYLER [1966].

Tribes based on the fundamental t-norm T_∞ were introduced in [BUTNARIU 1983a] where they were called *σ-algebras of fuzzy sets.* The slightly more special concepts of *fuzzy σ-algebras* based on T_0 and *T-fuzzy σ-algebras* based on arbitrary t-norms T [KLEMENT 1982a] were studied in [KLEMENT 1980a]. The triangular norm–based tribes discussed in this section are generalizations of this concept. Different generalizations of the concept of σ-algebra of fuzzy sets were studied by SCHMIDT [1985].

2.2 Let T be a t-norm and S be its corresponding t-conorm. We extend T and S to $[0,1]^X$ pointwise, i.e.,

$$(A\,\mathbf{T}\,B)(x) = T(A(x), B(x)) \tag{2.1}$$

and

$$(A\,\mathbf{S}\,B)(x) = S(A(x), B(x)). \tag{2.2}$$

These operations may be considered as *intersection* and *union* of fuzzy sets, respectively. Also, *finite* and *countable intersections*

$$\mathop{\mathbf{T}}_{i=1}^{n} A_i, \qquad \mathop{\mathbf{T}}_{n=1}^{\infty} A_i,$$

as well as finite and countable unions

$$\mathop{\mathbf{S}}_{i=1}^{n} A_i, \qquad \mathop{\mathbf{S}}_{n=1}^{\infty} A_i,$$

of fuzzy sets are defined in the straightforward way. They satisfy the De Morgan laws

$$(\mathop{\mathbf{T}}_{n=1}^{\infty} A_n)^{\mathrm{c}} = \mathop{\mathbf{S}}_{n=1}^{\infty} A_n^{\mathrm{c}}, \qquad (\mathop{\mathbf{S}}_{n=1}^{\infty} A_n)^{\mathrm{c}} = \mathop{\mathbf{T}}_{n=1}^{\infty} A_n^{\mathrm{c}} \tag{2.3}$$

where the *complement* A^c is defined by

$$A^c(x) = 1 - A(x). \tag{2.4}$$

Restricted to crisp sets, i.e., characteristic functions, these operations coincide with intersection, union and complement, respectively, regardless which t-norm and t-conorm are considered. The class $[0,1]^X$ of all fuzzy subsets of X, together with either one of the operations $\mathbf{T}$ and $\mathbf{S}$, forms a partially ordered commutative semigroup having $\emptyset$ as smallest (and null) element and X as largest (and unit) element. However, $[0,1]^X$ provided with the operations $\mathbf{T}$, $\mathbf{S}$ and the complement ^{c} is not a Boolean algebra. It is not even a lattice, except if $T = T_0$ and $S = S_0$. In general, $\mathbf{T}$ and $\mathbf{S}$ are not distributive with respect to each other, and $A\,\mathbf{T}\,A^c$ may be different from $\emptyset$ and $A\,\mathbf{S}\,A^c$ may be different from X.

2.3 Definition. Let T be a t-norm. A subfamily $\mathscr{C}$ of $[0,1]^X$ containing $\emptyset$ and being closed under the operation $\mathbf{T}$ and under complementation will be called a T-*clan*.

Obviously, by the duality of $\mathbf{T}$ and $\mathbf{S}$, the closedness with respect to $\mathbf{T}$ can be replaced by the closedness with respect to $\mathbf{S}$ in the definition above.

2.4 Example.

(i) Since we identify crisp subsets of X with their characteristic functions, any algebra of subsets of X is a T-clan with respect to each t-norm T.

(ii) For each $n \in \mathbb{N}$ the family $\mathscr{C}_n(X) = \{0, \frac{1}{n}, \frac{2}{n}, \ldots, \frac{n-1}{n}, 1\}^X$ is a T-clan for $T = T_0$ and also for $T = T_\infty$, but not with respect to any other fundamental t-norm.

(iii) If the t-norm T is a continuous or measurable function, and if X is a topological or measurable space, then the family of all continuous, or measurable, fuzzy subsets of X is a T-clan.

2.5 Definition. A T-clan $\mathscr{T}$ which is also closed under countable intersections, i.e., has the property

$$\mathop{\mathbf{T}}_{n=1}^{\infty} A_n \in \mathscr{T} \quad \text{for all sequences } (A_n)_{n\in\mathbb{N}} \subseteq \mathscr{T} \tag{2.5}$$

will be called a T-*tribe*. A pair $(X, \mathscr{T})$, where X is a set and $\mathscr{T}$ is a T-tribe, will be called T-*measurable space*.

Obviously, not each T-clan is a T-tribe. For instance, the T_∞-clan $([0,1] \cap \mathbb{Q})^X$ is not a T_∞-tribe. Each T-tribe is closed under countable *unions*.

2.6 Example.

(i) Each σ-algebra of crisp subsets of X is a T-tribe with respect to any t-norm T.

(ii) Given a σ-algebra $\mathscr{A}$ of crisp subsets of X, the family $\mathscr{A}^{\wedge}$ of all $\mathscr{A}$-measurable fuzzy subsets of X is a T-tribe with respect to any Borel measurable t-norm T.

(iii) Given a T-tribe $\mathscr{T}$, the set $\mathscr{T}^{\vee}$ of all characteristic functions contained in $\mathscr{T}$ is a σ-algebra and hence a T-tribe.

(iv) [KLEMENT 1980a] The family $\mathscr{T}$ consisting of all fuzzy subsets of $X = [0,1]$ which are either constant or have all their values in the interval $[\frac{1}{3}, \frac{2}{3}]$ is a T-tribe for $T = W$ (see (1.16)) and for $T = T_0$, but it is not a T-tribe for $T = T_t$ with $t \in]0, \infty]$. It is interesting to note that *there is no σ-algebra $\mathscr{A}$ of crisp subsets of X such that $\mathscr{T} = \mathscr{A}^{\wedge}$.*

(v) [KLEMENT 1982a] If in (iv) we additionally require that all elements of $\mathscr{T}$ be continuous, then $\mathscr{T}$ is a W-tribe but not a T_0-tribe. Indeed, the sequence of fuzzy sets $(A_n)_{n\in\mathbb{N}}$ given by

$$A_n(x) = \begin{cases} \frac{1}{3}(nx+1) & \text{if } x \le \frac{1}{n}, \\ \frac{2}{3} & \text{if } x > \frac{1}{n}, \end{cases}$$

is contained in the W-tribe $\mathscr{T}$, but the pointwise limit of this sequence, that is, the fuzzy set A given by

$$A(x) = \begin{cases} \frac{1}{3} & \text{if } x = 0, \\ \frac{2}{3} & \text{if } x > 0, \end{cases}$$

does not belong to $\mathscr{T}$.

(vi) Consider a nonempty, crisp subset Y of X such that $Y \neq X$. The family $\mathscr{T}$ of fuzzy subsets of X which are constant on Y and assume only values 0 and 1 outside of Y is a T-tribe with respect to any t-norm T, *but it does not contain any nontrivial constant fuzzy sets.*

2.7 Proposition.

(i) For all $s \in]0, \infty[$ we have: each T_s-tribe is a T_∞-tribe.

(ii) Each T_∞-tribe is a T_0 –tribe.

PROOF: In order to show the validity of (i), assume that $s \in]0, \infty[$. The proof is carried out in several steps.

(a) First we prove that for all $A, B \in \mathcal{T}$ there exists a $C \in \mathcal{T}$ such that $A \, \mathbf{S}_\infty \, B = A \, \mathbf{S}_s \, C$. Let A and B be two fuzzy sets in the T_s-tribe $\mathcal{T}$. Denote

$$\begin{aligned} A_1 &= A, & B_1 &= B, \\ A_{n+1} &= A_n \, \mathbf{S}_s \, B_n, & B_{n+1} &= A_n \, \mathbf{T}_s \, B_n. \end{aligned} \tag{2.6}$$

The sequence $(A_n)_{n \in \mathbb{N}}$ is nondecreasing, the sequence $(B_n)_{n \in \mathbb{N}}$ is nonincreasing, and both sequences are contained in $\mathcal{T}$. Since the pair (T_s, S_s) satisfies (1.25), it follows by induction that for all $n \in \mathbb{N}$ we have

$$A_n + B_n = A + B. \tag{2.7}$$

Claim 1. For each $a \in [0,1[$ there exists a number $c \in [0,1[$ such that for all b in $[0,a]$ we have

$$T_s(a, T_s(a,b)) \leq c \cdot T_s(a,b). \tag{2.8}$$

Indeed, from Proposition 1.11 we have $T_s \leq T_1$ whenever $s \geq 1$, in which case we may choose $c = a$. For $s < 1$ define the number

$$c = \frac{s^a - 1}{s - 1},$$

which clearly belongs to $[0,1[$. Then for each $b \in [0,a]$ the inequality (2.8) is equivalent to

$$\ln[1 + c^2(s^b - 1)] \geq c \cdot \ln[1 + c(s^b - 1)].$$

The expansion of the logarithms into a power series leads to

$$\sum_{i=1}^{\infty} (-1)^{i-1} \cdot \frac{c^{2i}(s^b - 1)^i}{i} \geq \sum_{i=1}^{\infty} (-1)^{i-1} \cdot \frac{c^{i+1}(s^b - 1)^i}{i}$$

which is equivalent to $c^{i-1} \leq 1$. This is true for all $i \in \mathbb{N}$ and it follows that (2.8) is valid for all $s \in]0, \infty[$.

Claim 2. If $C_1 = A_1$ and $C_{n+1} = B_n$ for all $n \in \mathbb{N}$, then we have

$$A \, \mathbf{S}_\infty \, B = \mathop{\mathbf{S}_s}_{n=1}^{\infty} C_n. \tag{2.9}$$

To prove this choose an arbitrary $x \in X$ and put $a = (A \, \mathbf{S}_\infty \, B)(x)$. If $a < 1$, then $A_n(x) \leq a < 1$ for all $n \in \mathbb{N}$ because of (2.7). Choose $c \in [0, a[$ such that (2.8) is satisfied. Then, by the monotonicity of the t-norm T_s we get

$$B_n(x) \leq \begin{cases} c^{n-2} \cdot a & \text{if } s < 1, \\ a^n & \text{if } s \geq 1, \end{cases}$$

for each $n \geq 2$. Since a and c are both in $[0,1[$, it follows that $\lim_{n\to\infty} B_n(x) = 0$ and, because of (2.7), we have

$$\lim_{n\to\infty} = A_n(x) = (A\,\mathbf{S}_\infty\, B)(x) \tag{2.10}$$

which is exactly (2.9) by virtue of (2.6). If $a = 1$ assume that (2.10) does not hold. Then $\lim_{n\to\infty} A_n(x) < 1$, and this means that there exists a number $d \in [0,1[$ such that $A_n(x) \leq d < 1$ for all $n \in \mathbb{N}$. Using arguments analogous to those above, with d instead of a, we deduce that $\lim_{n\to\infty} A_n(x) \geq 1$, contradicting our assumption. Hence, (2.9) is always true. Putting

$$C = \mathop{\mathbf{S}_s}_{n=2}^{\infty} C_n$$

the proof of part (a) is complete. This also shows that the T_s-tribe $\mathscr{T}$ is necessarily a T_∞-clan.

(b) Let $(D_n)_{n\in\mathbb{N}}$ be a sequence in $\mathscr{T}$. There exists a sequence $(E_n)_{n\in\mathbb{N}}$ in $\mathscr{T}$ such that for each $n \in \mathbb{N}$ we have

$$\mathop{\mathbf{S}_\infty}_{j=1}^{n} D_j = \mathop{\mathbf{S}_s}_{k=1}^{n} E_k. \tag{2.11}$$

For $n = 2$ this follows from part (a). Suppose we have proved (2.11) for n. Then we get, again using part (a),

$$\begin{aligned}
\mathop{\mathbf{S}_\infty}_{j=1}^{n+1} D_j &= (\mathop{\mathbf{S}_\infty}_{j=1}^{n} D_j)\,\mathbf{S}_\infty\, D_{n+1} \\
&= (\mathop{\mathbf{S}_s}_{j=1}^{n} E_j)\,\mathbf{S}_\infty\, D_{n+1} \\
&= (\mathop{\mathbf{S}_s}_{j=1}^{n} E_j)\,\mathbf{S}_s\, E_{n+1} \\
&= \mathop{\mathbf{S}_s}_{j=1}^{n+1} E_j.
\end{aligned}$$

Now, because of (2.11), we obtain

$$\mathop{\mathbf{S}_\infty}_{n=1}^{\infty} D_n = \lim_{n\to\infty} \mathop{\mathbf{S}_\infty}_{j=1}^{n} D_j = \lim_{n\to\infty} \mathop{\mathbf{S}_s}_{j=1}^{n} E_j = \mathop{\mathbf{S}_s}_{n=1}^{\infty} E_n,$$

the latter fuzzy set belonging to $\mathscr{T}$. Hence, $\mathscr{T}$ is a T_∞-tribe, and (i) is proven.

Now let us show (ii). In order to see that a T_∞-tribe $\mathscr{T}$ is always a T_0-clan note that for any two fuzzy sets A and B we always have

$$A\,\mathbf{T}_0\,B = (A\,\mathbf{S}_\infty\,B^c)\,\mathbf{T}_\infty\,B.$$

Moreover, $\mathscr{T}$ is even a T_0-tribe. If $(A_n)_{n\in\mathbb{N}}$ is an increasing sequence in $\mathscr{T}$, put $A_0 = \emptyset$ and for each $n \in \mathbb{N}$ define $B_n = A_n\,\mathbf{T}_\infty\,A_{n-1}^c$. Observe that for $n \in \mathbb{N}$

$$A_n = \mathop{\mathbf{S}_\infty}_{j=1}^{n} B_j,$$

and therefore

$$\mathop{\mathbf{S}_0}_{n=1}^{\infty} A_n = \lim_{n\to\infty} A_n = \mathop{\mathbf{S}_\infty}_{n=1}^{\infty} B_n \in \mathscr{T},$$

thus completing the proof. □

2.8 Remark. Theorem 2.7 shows that T-tribes based on fundamental t-norms T are implicitly σ-complete lattices with respect to the pointwise order and that they are closed sets with respect to the weak topology on the unit cube $[0,1]^X$. Moreover, viewed as T_∞-tribes, the fundamental norm-based T_s-tribes with $s \in]0,\infty]$ are implicitly clans in the sense of WYLER [1966], the substraction being defined by $A \ominus B = A\,\mathbf{T}_\infty\,B^c$ (see also [SCHMIDT 1985]).

3 Tribes Based on Fundamental t-Norms

3.1 We have already observed that each T-tribe $\mathscr{T}$ on X contains a σ-algebra $\mathscr{T}^\vee$ of crisp subsets of X, and that the set $\mathscr{T}^\circ$ of all $\mathscr{T}^\vee$-measurable functions $X \longrightarrow [0,1]$ is a T-tribe whenever the t-norm T is Borel measurable (see Example 2.6 (ii)). Now we want to study the relationship between $\mathscr{T}$ and $\mathscr{T}^\circ$. In particular, it is interesting to know under which conditions they coincide. In such a case, the T-tribe $\mathscr{T}$ is called *generated*. For an example of a nongenerated T-tribe see Example 2.6 (iv)).

3.2 The proof of the following result is essentially due to Z. ARTSTEIN [private communication]. An alternative proof of the same result appears in [BUTNARIU & KLEMENT 1991].

Theorem. *For each $s > 0$ and for each T_s-tribe $\mathscr{T}$ we have*

$$\mathscr{T} \subseteq \mathscr{T}^\circ. \tag{3.1}$$

PROOF: Fixing an $A \in \mathscr{T}$ and denoting for each $\alpha \in [0,1]$

$$A_\alpha = \{x \in X \mid A(x) \geq \alpha\}, \tag{3.2}$$

then we have to show that $A_\alpha \in \mathscr{T}^\vee$. For $\alpha = 0$ this is trivial. Assume $\alpha = 1$. Then for each constant sequence $(B_n)_{n\in\mathbb{N}}$ of fuzzy sets, i.e., $B_n = B$ for all $n \in \mathbb{N}$, we get from Proposition 1.12

$$B(x) > 0 \quad \text{if and only if} \quad \mathop{\mathbf{S}_s}_{n=1}^{\infty} B_n(x) = 1. \tag{3.3}$$

Putting $B = A^c$ this shows $A_1 \in \mathscr{T}^\vee$. Assume $\alpha \in]0,1[$. Then there exists an increasing sequence $(\alpha_n)_{n\in\mathbb{N}}$ of positive rational numbers converging to α, and we have

$$A_\alpha = \bigcap_{n=1}^{\infty} A_{\alpha_n}.$$

Thus it suffices to show that $A_\alpha \in \mathscr{T}^\vee$ for each $\alpha \in]0,1[$ which can be represented in finite binary form

$$\alpha = \sum_{i=1}^{k} \frac{\alpha_i}{2^i} \tag{3.4}$$

with $k \in \mathbb{N}$ and $\alpha_i \in \{0,1\}$ for each $i \in \{1,2,\ldots,k\}$. We proceed by induction upon the positive integer k. If $k = 1$, then $\alpha_1 = 1$ and $\alpha = \frac{1}{2}$ and we have

$$A_\alpha = \{x \in X \mid (A\,\mathbf{T}_\infty\,A)(x) = 1\} = (A\,\mathbf{T}_\infty\,A)_1.$$

By Proposition 2.7, $A\,\mathbf{T}_\infty\,A \in \mathscr{T}$. Therefore, $A_\alpha = (A\,\mathbf{T}_\infty\,A)_1$ belongs to $\mathscr{T}^\vee$. Assume that for each $A \in \mathscr{T}$ and for each $\alpha \in]0,1[$ of the form (3.4) with $k \leq n$ we have $A_\alpha \in \mathscr{T}^\vee$. Suppose that

$$\alpha = \sum_{i=1}^{n+1} \frac{\alpha_i}{2^i} \tag{3.5}$$

with $\alpha_i \in \{0,1\}$ for $i \in \{1,2,\ldots,n+1\}$ and $\alpha_{n+1} \neq 0$. Then $\alpha = \frac{\beta}{2}$ where

$$\beta = \sum_{i=1}^{n+1} \frac{\alpha_i}{2^{i-1}} \tag{3.6}$$

If $\beta \leq 1$, then $A_\alpha = (A\,T_\infty\,A)_\beta \in \mathscr{T}^\vee$ by the inductive assumption. If $\beta > 1$ then $\alpha > \frac{1}{2}$, and we may write $\alpha = \frac{\gamma+1}{2}$ with

$$\gamma = \sum_{i=1}^{n} \frac{\alpha_{i+1}}{2^i} \in [0,1]. \tag{3.7}$$

Thus we have $A_\alpha = (A\,\mathbf{T}_\infty\, A)_\gamma$, and this belongs to $\mathcal{T}^\vee$ by Proposition 2.7 and the inductive assumption. □

3.3 In general, a T_s-tribe $\mathcal{T}$ is not necessarily generated (i.e., $\mathcal{T}$ may be different from $\mathcal{T}^\circ$) even if T_s is a fundamental t-norm with $s > 0$ (see Example 2.6 (iv)). However, we have the following result:

Proposition. *If $s \in]0, \infty]$, then a T_s-tribe on X is generated if and only if it contains all the constant fuzzy subsets of X.*

PROOF: Necessity is obvious. Conversely, assume that the T_s-tribe $\mathcal{T}$ contains all constant fuzzy subsets of X. We have to show that each $A \in \mathcal{T}^\circ$ is contained in $\mathcal{T}$ (see Theorem 3.2). Define the sequence $(U_n)_{n\in\mathbb{N}}$ of fuzzy sets by

$$U_n = \mathop{\mathbf{S}_\infty}_{k=1}^{2^n-1} (A_k^{[n]}\,\mathbf{T}_\infty\, V_k^{[n]}), \tag{3.8}$$

where

$$V_k^{[n]} = \begin{cases} \{x \in X \mid \frac{2k-1}{2^n} \le A(x) < \frac{2k}{2^n}\} & \text{if } k < 2^n - 1, \\ \{x \in X \mid \frac{2^n-1}{2^n} \le A(x)\} & \text{if } k = 2^n - 1, \end{cases} \tag{3.9}$$

and the constant fuzzy set $A_k^{[n]}$ is given by $A_k^{[n]}(x) = \alpha_k^{[n]}$, with the number $\alpha_k^{[n]}$ chosen such that

$$S_s\left(\tfrac{2k-2}{2^n}, \alpha_k^{[n]}\right) = \frac{2k-1}{2^n}. \tag{3.10}$$

Note that this choice of $\alpha_k^{[n]}$ is possible since the function

$$S_s\left(\tfrac{2k-2}{2^n}, \cdot\right) : [0,1] \longrightarrow \left[\tfrac{2k-2}{2^n}, 1\right], \quad x \longmapsto S_s\left(\tfrac{2k-2}{2^n}, x\right)$$

is a surjection. Because of the $\mathcal{T}^\vee$-measurability of A, each $V_k^{[n]}$ is contained in $\mathcal{T}^\vee$ and hence in $\mathcal{T}$. Since $\mathcal{T}$ contains the constant fuzzy sets $A_k^{[n]}$ and since it is a T_∞-clan (cf. Proposition 2.7), it follows that all $U_n \in \mathcal{T}$. Now it is a matter of computation to see that

$$\mathop{\mathbf{S}_s}_{i=1}^{n} U_i = \sum_{k=1}^{2^n} \frac{k-1}{2^n} \cdot W_k^{[n]}, \tag{3.11}$$

where

$$W_k^{[n]} = \begin{cases} \{x \in X \mid \frac{k-1}{2^n} \le A(x) < \frac{k}{2^n}\} & \text{if } k < 2^n, \\ \{x \in X \mid \frac{2^n-1}{2^n} \le A(x)\} & \text{if } k = 2^n. \end{cases}$$

Since the functions on the right hand side of (3.11) are convergent to A if $n \to \infty$, we get

$$A = \mathop{\mathbf{S}_s}_{n=1}^{\infty} U_n.$$

□

3.4 Remark.

(i) According to Proposition 2.7, in Propositon 3.3 the condition "the T-tribe $\mathcal{T}$ contains all constant fuzzy subsets of X" can be replaced by the condition "there exists an integer $k \geq 2$ such that all constant fuzzy sets of the form $A_n(x) = 1/k^n$, where $n \in \mathbb{N}$, belong to $\mathcal{T}$ ".

(ii) Theorem 3.2 and Proposition 3.3 show that a generated T_s-tribe $\mathcal{T}$ is necessarily contained in the nonnegative wedge of the unit ball of the space $L_1(X, \mathcal{T}^\vee)$, and that nongenerated T_s-tribes are special subsets of this space. These facts will be used in Chapter V.

(iii) In [MESIAR, to appear] it is shown that for $s \in]0, \infty[$ each T_s-tribe $\mathcal{T}$ on a *countable* set X is a *semigenerated* tribe, i.e., there are disjoint crisp subsets $Y, Z \subseteq X$ with $X = Y \cup Z$ such that $\mathcal{T}/_Y$, the restriction of $\mathcal{T}$ to Y, is a generated tribe and $\mathcal{T}/_Z$ is a σ-algebra of crisp subsets of Z. Consequently, if $\mathcal{T}$ is a T_s-tribe on a countable set X for some $s \in]0, \infty[$, it is a T-tribe for any measurable t-norm T. For an example of a semigenerated tribe $\mathcal{T}$ which is not generated see Example 2.6(vi).

Chapter II

Triangular Norm–Based Measures

4 Disjointness with Respect to t-Norms

4.1 *Triangular norm–based measures*, or simply T*-measures*, are special real valuations defined on T-tribes. They attracted the interest of scientists because of their applications in fields like mathematical statistics (see [DVORETZKY et al. 1950, 1951]), pattern recognition ([ZADEH 1968], [SUGENO 1974]), possibility and plausibility theory ([ZADEH 1978], [HÖHLE & KLEMENT 1984], etc.) and multivalued logics ([ROSE & ROSSER 1958]). In this book T-measures will be used in order to identify classes of games for which *Aumann–Shapley values* (see Chapter IV) exist. Among the basic tools employed in this respect are *integral representations* of T-measures which form the main topic of this chapter.

In order to classify the T-measures and to distinguish between T-measures and other kinds of valuations which can be defined on T-tribes (see, for instance, [SCHMIDT 1982]), we first introduce the concepts of T*-disjointness.*

Definition. Let X be a nonempty set, T a t-norm and S its corresponding t-conorm.

(i) A finite family $(A_1, A_2, \ldots, A_n)$ of fuzzy sets in $[0,1]^X$ is said to be T*-disjoint* if for each $k \in \{1, \ldots, n\}$

$$(\mathbf{S}_{j \neq k} A_j) \, \mathbf{T} \, A_k = \emptyset. \tag{4.1}$$

(ii) A countable family $(A_n)_{n \in \mathbb{N}} \subseteq [0,1]^X$ is called T-disjoint if, for each $n \in \mathbb{N}$, the finite family $(A_1, \ldots, A_n)$ is T-disjoint.

(iii) A sequence $(A_n)_{n \in \mathbb{N}} \subseteq [0,1]^X$ is called a T*-partition* of a fuzzy set A if it is

T-disjoint and if

$$A = \mathbf{S}_{n=1}^{\infty} A_n.$$

4.2 Remark.

(i) Each subfamily $(A_n)_{n\in I}$ of an at most countable T-disjoint family $(A_n)_{n\in J}$ is also T-disjoint. Obviously, it suffices to show this for finite index sets J and $I \subseteq J$. Indeed, for each $i \in I$ we get

$$\left(\mathbf{S}_{n\in I\setminus\{i\}} A_n\right) \mathbf{T}\, A_i \leq \Big(\mathbf{S}_{n\in J\setminus\{i\}} A_n\Big) T\, A_i = \emptyset.$$

(ii) The definition of the T-disjointness does not depend on the order in which the fuzzy sets $(A_n)_{n\in\mathbb{N}}$ are numbered, i.e., if π is a permutation of $\mathbb{N}$ and the countable family $(A_n)_{n\in\mathbb{N}}$ is T-disjoint, then the permuted family $\left(A_{\pi(n)}\right)_{n\in\mathbb{N}}$ is T-disjoint too.

4.3 Different t-norms may lead to different concepts of disjointness. However, for some classes of t-norms the corresponding disjointness concepts do not depend on the choice of the t-norm in that class.

Example. Let $(A_n)_{n\in\mathbb{N}}$ be a countable family of fuzzy subsets of X.

(i) If all A_n are crisp subsets of X, then for each t-norm T the T-disjointness is equivalent to the pairwise disjointness of the crisp sets.

(ii) For each t-norm T, the T-disjointness implies pairwise T-disjointness according to 4.2(i), but the converse is not generally true: if we take the family of constant fuzzy sets (A_1, A_2, A_3) given by $A_1(x) = A_2(x) = A_3(x) = 0.5$, then A_1, A_2, A_3 are pairwise T_∞-disjoint, but the family (A_1, A_2, A_3) is not T_∞-disjoint.

(iii) For $s \in [0, \infty[$ we get: $(A_n)_{n\in\mathbb{N}}$ is T_s-disjoint if and only if each element $x \in X$ is "contained" in at most one A_k (i.e., if and only if $A_k(x) > 0$ for at most one $k \in \mathbb{N}$).

(iv) For $T = W$ (see (1.16)) T-disjointness of $(A_n)_{n\in\mathbb{N}}$ means that for each $x \in X$ exactly one of the following conditions holds:

  (a) There is exactly one $k \in \mathbb{N}$ such that $A_k(x) = 1$, and $A_n(x) = 0$ for all $n \neq k$.

  (b) There are at most two indices $k, l \in \mathbb{N}$ such that $0 < A_k(x), A_l(x) < 1$ and $A_n(x) = 0$ for $n \neq k$ and $n \neq l$.

4.4 Lemma. *Let T be a t-norm and S be its corresponding t-conorm such that the functional equation (1.25) holds, and let $(A_1, A_2 \ldots, A_n)$ be a finite family of fuzzy subsets of X. Then for each $n \geq 2$ the following are equivalent:*

(i) the family $(A_1, A_2, \ldots, A_n)$ is T-disjoint;

(ii) for all $k \in \{2, \ldots, n\}$ we have

$$(\mathop{\mathbf{S}}_{i=1}^{k-1} A_i)\,\mathbf{T}\, A_k = \emptyset;$$

(iii) for all $k \in \{2, \ldots, n\}$ we have

$$\mathop{\mathbf{S}}_{i=1}^{k} A_i = \sum_{i=1}^{k} A_i;$$

(iv) for each subset $I \subseteq \{1, 2, \ldots, n\}$ containing at least $n-1$ elements we have

$$\mathop{\mathbf{S}}_{i \in I} A_i = \sum_{i \in I} A_i.$$

PROOF: Implication (i) $\Rightarrow$ (ii) is an immediate consequence of Remark 4.2(i). To verify (ii) $\Rightarrow$ (iii) we use (1.25) and get

$$(\mathop{\mathbf{S}}_{i=1}^{k-1} A_i)\,\mathbf{T}\, A_k + (\mathop{\mathbf{S}}_{i=1}^{k-1} A_i)\,\mathbf{S}\, A_k = \mathop{\mathbf{S}}_{i=1}^{k-1} A_i + A_k \tag{4.2}$$

which implies

$$(\mathop{\mathbf{S}}_{i=1}^{k} A_i) = \mathop{\mathbf{S}}_{i=1}^{k-1} A_i + A_k.$$

Repeating this $(k-2)$ times gives the desired result.

Now let us show (iii) $\Rightarrow$ (iv). In the cases $n = 2$ or $I = \{1, 2, \ldots, n\}$ or $I = \{1, 2, \ldots, n-1\}$, there is nothing to prove. Otherwise, observe that for $k \in \{2, \ldots, n\}$ we have (4.2) which implies

$$(\mathop{\mathbf{S}}_{i=1}^{k-1} A_i)\,\mathbf{T}\, A_k = \emptyset. \tag{4.3}$$

Now, for $k \in \{2, \ldots, n+1\}$ and $l \leq k-1$ define $I_{k,l} = \{1, \ldots, k-1\} \setminus \{l\}$. Then (4.2) and the monotonicity of S and T together imply

$$(\mathop{\mathbf{S}}_{i \in I_{k,l}} A_i)\,\mathbf{T}\, A_k = \emptyset. \tag{4.4}$$

Since from (1.25) we have

$$(\mathbf{S}_{i\in I_{k,l}} A_i)\,\mathbf{T}\,A_k + (\mathbf{S}_{i\in I_{k,l}} A_i)\,\mathbf{S}\,A_k = (\mathbf{S}_{i\in I_{k,l}} A_i) + A_k$$

and (4.4) holds, we get

$$\mathbf{S}_{i\in I_{k+1,l}} A_i = (\mathbf{S}_{i\in I_{k,l}} A_i) + A_k. \tag{4.5}$$

Now, put $k = n$ and $l \leq n-1$. In this case, if $l = n-1$, we obtain the desired result from (4.5). If $l < n-1$, compute $\mathbf{S}_{i\in I_{k,l}} A_i$ using (4.5) and insert it into (4.5) again. Continuing until $l = k-1$ we again obtain the desired result. In order to show (iv) $\Rightarrow$ (i), choose $k \in \{1,2,\ldots,n\}$ and put $I_k = \{1,2,\ldots,n\} \setminus \{k\}$. Then (1.25) yields

$$(\mathbf{S}_{i\in I_k} A_i)\,\mathbf{T}\,A_k + (\mathbf{S}_{i\in I_k} A_i)\,\mathbf{S}\,A_k = \mathbf{S}_{i\in I_k} A_i + A_k$$

which immediately implies the T-disjointness. □

4.5 Corollary. *Let T be a t-norm and S its corresponding t-conorm such that the functional equation (1.25) holds, and let $(A_n)_{n\in\mathbb{N}}$ be a countable family of fuzzy subsets of X. Then the following are equivalent:*

(i) the family $(A_n)_{n\in\mathbb{N}}$ is T-disjoint;

(ii) for all $k \geq 2$ we have

$$(\mathbf{S}_{i=1}^{k-1} A_i)\,\mathbf{T}\,A_k = \emptyset;$$

(iii) for all $k \geq 2$ we have

$$\mathbf{S}_{i=1}^{k} A_i = \sum_{i=1}^{k} A_i;$$

(iv) for each finite subset $I \subseteq \mathbb{N}$ we have

$$\mathbf{S}_{i\in I} A_i = \sum_{i\in I} A_i.$$

4.6 Remark. Let $(A_n)_{n\in J}$ be an at most countable family of fuzzy sets.

(i) $(A_n)_{n\in J}$ is T_∞-disjoint if and only if $\sum_{n\in J} A_n \leq 1$.

(ii) From Lemma 4.4 and Corollary 4.5 we know that if T and its corresponding t-conorm S satisfy (1.25) and if $(A_n)_{n\in J}$ is T-disjoint, then necessarily $\sum_{n\in J} A_n \leq 1$. However, the converse is not generally true (see Example 4.3(iii)).

(iii) The requirement that T and S satisfy (1.25) cannot be dropped in Lemma 4.4 and in Corollary 4.5. For instance, for the t-norm W and its corresponding t-conorm V (see Example 3.3(iv)) the conditions (i), (ii) and (iii) are no longer equivalent.

5 Triangular Norm-Based Measures and their Representation by Markov Kernels

5.1 Throughout this section X denotes a nonempty set, T a t-norm, and S its corresponding t-conorm. For a subset $\mathscr{T} \subseteq [0,1]^X$ we consider functions $\boldsymbol{m} : \mathscr{T} \longrightarrow [-\infty, +\infty]$ which assume *at most one of the values* $-\infty$ and $+\infty$.

Definition. Let $\mathscr{T}$ be a T-clan.

(i) *A* function $\boldsymbol{m} : \mathscr{T} \longrightarrow [-\infty, +\infty]$ is called a *T-valuation (on $\mathscr{T}$)* if it satisfies the following conditions:

$$\boldsymbol{m}(\emptyset) = 0. \tag{5.1}$$

$$\boldsymbol{m}(A \mathbf{T} B) + \boldsymbol{m}(A \mathbf{S} B) = \boldsymbol{m}(A) + \boldsymbol{m}(B). \tag{5.2}$$

(ii) The function $\boldsymbol{m} : \mathscr{T} \longrightarrow [-\infty, +\infty]$ is said to be *finitely T-additive* if it satisfies (5.1) and

$$\boldsymbol{m}(A \mathbf{S} B) = \boldsymbol{m}(A) + \boldsymbol{m}(B) \quad \text{whenever} \quad A \mathbf{T} B = \emptyset \tag{5.3}$$

(iii) A function $\boldsymbol{m} : \mathscr{T} \longrightarrow [-\infty, +\infty]$ is said to be *monotone* if

$$\boldsymbol{m}(A) \leq \boldsymbol{m}(B) \quad \text{whenever} \quad A \leq B. \tag{5.4}$$

5.2 Remark.

(i) If $\boldsymbol{m} : \mathscr{T} \longrightarrow [-\infty, +\infty]$ is a T-valuation on the T-clan $\mathscr{T}$, then $\boldsymbol{m}$ is also finitely T-additive, the converse not being generally true: if, for instance, $\mathscr{T}$ consists of all the constant functions in $[0,1]^X$ and if $s \in [0, \infty[$, then because of the absence of any nontrivial T_s-disjoint elements in the T_s-clan $\mathscr{T}$, each function $\boldsymbol{m} : \mathscr{T} \longrightarrow [-\infty, +\infty]$ which satisfies (5.1) is finitely T_s-additive without being necessarily a T_s-valuation. This shows that T-valuations are particular additive functions in the sense of SCHMIDT [1982, p. 558] and, consequently, if they are finite, they can be represented as differences of monotone T-additive functions (cf. [SCHMIDT 1982, Theorem 2.2]). However, *it is not possible to conclude directly that the T-valuations have Jordan decompositions.*

(ii) If $\mathscr{T}$ is a T_∞-clan and $\boldsymbol{m}$ is a finite T_∞-additive function on $\mathscr{T}$, then $\boldsymbol{m}$ is also a T_∞-valuation.

(iii) If $\mathscr{T}$ is a T-clan consisting of crisp subsets of X only, then the finite T-additive functions are Q-valuations for any t-norm Q.

5.3 Definition.

(i) Let $\mathscr{T}$ be a T-clan. A function $\boldsymbol{m} : \mathscr{T} \longrightarrow [-\infty, +\infty]$ is called a *T-measure* if it is a T-valuation and continuous from the left in the following sense:

$$\lim_{n\to\infty} \boldsymbol{m}(A_n) = \boldsymbol{m}(A) \quad \text{whenever} \quad (A_n)_{n\in\mathbb{N}} \subseteq \mathscr{T}, (A_n)_{n\in\mathbb{N}} \nearrow A \text{ and } A \in \mathscr{T}. \tag{5.5}$$

(ii) If a T-measure $\boldsymbol{m}$ on $\mathscr{T}$ assumes only values in [0,1] and if m(X) = 1, we shall call $\boldsymbol{m}$ a *probability T-measure*.

(iii) Let $\mathscr{T}$ be a T-tribe. The function $\boldsymbol{m} : \mathscr{T} \longrightarrow [-\infty, +\infty]$ is said to be *countably T-additive* if it satisfies (5.1) and if for each sequence $(A_n)_{n\in\mathbb{N}} \subseteq \mathscr{T}$ we have

$$\boldsymbol{m}(\mathbf{S}_{i=1}^{\infty} A_n) = \sum_{i=1}^{\infty} \boldsymbol{m}(A_n) \quad \text{whenever} \quad (A_n)_{n\in\mathbb{N}} \text{ is } T\text{-disjoint} \tag{5.6}$$

5.4 Remark. It is clear that countably T-additive functions are always finitely T-additive. Moreover, T-measures on T-tribes are countably T-additive since if we take a T-disjoint sequence $(A_n)_{n\in\mathbb{N}}$ in the T-tribe $\mathscr{T}$ and define $B_n = \mathbf{S}_{i=1}^{n} A_n$, then $(B_n)_{n\in\mathbb{N}}$ is a sequence in $\mathscr{T}$ which converges nondecreasingly to $\mathbf{S}_{n=1}^{\infty} A_n$ and, therefore,

$$\boldsymbol{m}(\mathbf{S}_{n=1}^{\infty} A_n) = \boldsymbol{m}(\lim_{n\to\infty} B_n) = \lim_{n\to\infty} \boldsymbol{m}(B_n) = \lim_{n\to\infty} \sum_{i=1}^{n} \boldsymbol{m}(A_n) = \sum_{n=1}^{\infty} \boldsymbol{m}(A_n).$$

In general, for arbitrary t-norms T and T-tribes $\mathscr{T}$, countable T-additivity *does not imply* the continuity from the left condition (5.5) and thus countably T-additive functions are not necessarily T-measures. For example, if $T = T_s$ with $s \in [0, \infty[$ and $\mathscr{T} = [0,1]^X$, then for each fixed $x_0 \in X$ the function $\boldsymbol{m} : \mathscr{T} \longrightarrow [-\infty, +\infty]$ defined by

$$\boldsymbol{m}(A) = \begin{cases} 1 & \text{if } A(x_0) = 1 \\ 0 & \text{if } A(x_0) < 1 \end{cases} \tag{5.7}$$

is countably T_s-additive, but it is not a T_s-measure since it does not satisfy (5.5).

However, it will be shown later (see Proposition 9.1) that if $T = T_\infty$, then countably T-additive functions are T-measures, that is, the countable T-additivity and the continuity from the left are equivalent in this case. The same is true when T is an arbitrary t-norm and $\mathscr{T}$ consists of crisp subsets of X only.

If $\mathscr{T}$ is a σ-algebra of crisp subsets of X, then each (σ-additive) measure on $\mathscr{T}$ is a T-measure with respect to any t-norm T. In this case, the countable T-additivity and the continuity from the left are equivalent too.

5.5 The following result will be used when we represent T_s-measures by Markov kernels.

Lemma. *If $\mathscr{T}$ is both a T-clan and a T_0-clan, then each T-valuation on $\mathscr{T}$ is also a T_0-valuation on $\mathscr{T}$.*

PROOF: Let A and B be in $\mathscr{T}$. Then we have

$$\begin{aligned} \boldsymbol{m}(A\,\mathbf{T}_0\,B) + \boldsymbol{m}(A\,\mathbf{S}_0\,B) &= \boldsymbol{m}((A\,\mathbf{T}_0\,B)\,\mathbf{T}\,(A\,\mathbf{S}_0\,B) + \boldsymbol{m}((A\,\mathbf{T}_0\,B)\,\mathbf{S}\,(A\,\mathbf{S}_0\,B)) \\ &= \boldsymbol{m}(A\,\mathbf{T}\,B) + \boldsymbol{m}(A\,\mathbf{S}\,B) \\ &= \boldsymbol{m}(A) + \boldsymbol{m}(B), \end{aligned}$$

showing that $\boldsymbol{m}$ is indeed a T_0-valuation. □

5.6 Remark.

(i) Lemma 5.5 shows that if $\boldsymbol{m}$ is a T-measure on $\mathscr{T}$ ($\mathscr{T}$ being both a T- and a T_0-clan), then $\boldsymbol{m}$ is also a T_0-measure.

(ii) If $\boldsymbol{m}$ is a T_s-measure on a T_s-tribe $\mathscr{T}$ with $s \in]0, \infty]$, then it is also a T_0-measure (cf. Theorem 2.7).

(iii) The converse of Lemma 5.5 does not generally hold: for instance, let $\mathscr{T}$ be the family of all Borel measurable fuzzy sets on $X = [0, 1]$. Then the function $\boldsymbol{m} : \mathscr{T} \longrightarrow [-\infty, +\infty]$ defined by

$$\boldsymbol{m}(A) = \int_{\{A>0\}} (1 + A(x))\,\mathrm{d}x \tag{5.8}$$

is a T_0-valuation (even a T_0-measure) but not a T_∞-valuation.

5.7 The T_0-measures play a fundamental role in the following exposition. In order to give an integral representation for them we need the notion of Markov–kernels (for more details see, e.g., [BAUER 1972]).

Definition. Let $(X, \mathscr{A})$ be a measurable space, $\mathscr{B}_0$ be the family of all Borel subsets of $[0,1]$, and $\mathscr{B}_1 = \mathscr{B}_0 \cap [0,1[$. A function

$$\mathfrak{K} : X \times \mathscr{B}_1 \longrightarrow \mathbb{R}, \quad (x, B) \longmapsto \mathfrak{K}(x, B)$$

is called an *$\mathscr{A}$-Markov–kernel* if it satisfies the following conditions:

(i) For each $x \in X$, the following function is a probability measure on $\mathscr{B}_1$:

$$\mathfrak{K}(x, \cdot) : \mathscr{B}_1 \longrightarrow \mathbb{R}, \quad B \longmapsto \mathfrak{K}(x, B).$$

(ii) For each $B \in B_1$, the following function is $\mathscr{A}$-measurable:

$$\mathfrak{K}(\cdot, B) : X \longrightarrow \mathbb{R}, \quad x \longmapsto \mathfrak{K}(x, B).$$

5.8 From Example 2.6(ii) we know that if $(X, \mathscr{A})$ is a measurable space, then the family $\mathscr{A}^\wedge$ of all $\mathscr{A}$-measurable functions from X to $[0,1]$ is a T_0-tribe. The next result, from [KLEMENT 1980b], shows that T_0-measures on $\mathscr{A}^\wedge$ can be represented as integrals of Markov–kernels in the following way:

Theorem. (First Representation Theorem). *Let $\mathscr{A}$ be a σ-algebra of crisp subsets of X and $\mathscr{A}^\wedge$ the family of all $\mathscr{A}$-measurable functions in $[0,1]^X$. If $\boldsymbol{m}$ is a finite monotone T_0-measure on $\mathscr{A}^\wedge$, then there exists a unique measure $\check{\boldsymbol{m}}$ on $\mathscr{A}$, namely, the restriction of $\boldsymbol{m}$ to $\mathscr{A}$, and an $\check{\boldsymbol{m}}$-almost everywhere uniquely determined $\mathscr{A}$-Markov–kernel $\mathfrak{K} : X \times \mathscr{B}_1 \longrightarrow \mathbb{R}$ such that for all $A \in \mathscr{A}$*

$$\boldsymbol{m}(A) = \int_X \mathfrak{K}(x, [0, A(x)[)\, d\check{\boldsymbol{m}}(x). \tag{5.9}$$

PROOF: The proof is carried out in several steps. First, observe that if (5.9) holds for some measure $\check{\boldsymbol{m}}$ on $\mathscr{A}$, then $\check{\boldsymbol{m}}$ has to be necessarily the restriction of $\boldsymbol{m}$ to $\mathscr{A}$ since for $M \in \mathscr{A}$ equality (5.9) gives

$$\boldsymbol{m}(M) = \int_M \mathfrak{K}(x, [0,1[)\, d\check{\boldsymbol{m}}(x) = \check{\boldsymbol{m}}(M). \tag{5.10}$$

Without loss of generality we may restrict our proof to the case when $\boldsymbol{m}(X) = 1$.

Claim 1. For each rational number $\alpha \in [0,1]$, the function $\boldsymbol{m}_\alpha : \mathscr{A} \longrightarrow [0,1]$, defined by

$$\boldsymbol{m}_\alpha(M) = \boldsymbol{m}(\alpha \cdot M), \tag{5.11}$$

is a nonnegative finite measure on $\mathscr{A}$, and it satisfies $\boldsymbol{m}_\alpha \leq \check{\boldsymbol{m}}$.

Obviously, $\boldsymbol{m}_\alpha(\emptyset) = 0$ and $\boldsymbol{m}_\alpha$ is finite and monotone because of the corresponding properties of $\boldsymbol{m}$. To prove σ-additivity of $\boldsymbol{m}_\alpha$, let $(A_n)_{n\in\mathbb{N}}$ be a sequence of mutually disjoint sets in $\mathscr{A}$. Then

$$\begin{aligned}
\boldsymbol{m}_\alpha(\bigcup_{n\in\mathbb{N}} A_n) &= \boldsymbol{m}\left[\sup_{n\in\mathbb{N}}(\alpha \cdot (\bigcup_{i=1}^{n} A_i))\right] \\
&= \sup_{n\in\mathbb{N}} \boldsymbol{m}(\alpha \cdot (\bigcup_{i=1}^{n} A_i)) \\
&= \sup_{n\in\mathbb{N}} \sum_{i=1}^{n} \boldsymbol{m}(\alpha \cdot A_i) \\
&= \sum_{n=1}^{\infty} \boldsymbol{m}(\alpha \cdot A_n).
\end{aligned}$$

The last equality follows from the monotonicity of $\boldsymbol{m}$. Therefore, Claim 1 is proved.

Claim 1 implies that each measure $\boldsymbol{m}_\alpha$ defined by (5.11) is $\check{\boldsymbol{m}}$-absolutely continuous. Hence, for each $\alpha \in [0,1] \cap \mathbb{Q}$ there exists a Radon–Nikodym derivative

$$f_\alpha = \frac{\mathrm{d}\boldsymbol{m}_\alpha}{\mathrm{d}\check{\boldsymbol{m}}}.$$

Since the Radon–Nikodym derivative of an $\check{\boldsymbol{m}}$-absolutely continuous measure is $\check{\boldsymbol{m}}$-a.e. unique, it follows that for all $\alpha \in]0,1[$

$$\begin{aligned}
f_0 &= 0 && \check{\boldsymbol{m}}\text{ -a. e.} \\
f_1 &= 1 && \check{\boldsymbol{m}}\text{ -a. e.} \\
f_\alpha &= \sup_{\beta\in\mathbb{Q}\cap[0,\alpha[} f_\beta && \check{\boldsymbol{m}}\text{ -a. e.}
\end{aligned} \tag{5.12}$$

Therefore, changing (if necessary) the values of the functions f_α on an $\check{\boldsymbol{m}}$-null set, we may assume that (5.12) holds everywhere on X, and we still have

$$\boldsymbol{m}_\alpha(M) = \int_M f_\alpha \,\mathrm{d}\check{\boldsymbol{m}}\,. \tag{5.13}$$

For an arbitrary (not necessarily rational) $\alpha \in [0,1]$ define

$$g_\alpha = \sup_{\beta\in\mathbb{Q}\cap[0,\alpha[} f_\beta \tag{5.14}$$

Since g_α is the supremum of a countable family of $\mathscr{A}$-measurable functions, it is also $\mathscr{A}$-measurable, and for each $M \in \mathscr{A}$ we again have

$$\boldsymbol{m}_\alpha(M) = \int_M g_\alpha \,\mathrm{d}\check{\boldsymbol{m}} \tag{5.15}$$

because of (5.12) and the Monotone Convergence Theorem. Furthermore, for each $x \in X$ the function $h_x : [0,1] \longrightarrow \mathbb{R}$ defined by

$$h_x(\alpha) = g_\alpha(x) \tag{5.16}$$

is a probability distribution function which determines a unique probability measure $\mathbf{q}_x$ on $([0,1[\,,\mathscr{B}_1)$ which, whenever $0 \leq \alpha < \beta \leq 1$, satisfies

$$\mathbf{q}_x([\alpha,\beta[\,) = h_x(\beta) - h_x(\alpha).$$

Claim 2. The function $\mathfrak{K} : X \times \mathscr{B}_1 \longrightarrow \mathbb{R}$ defined by

$$\mathfrak{K}(x,B) = \mathbf{q}_x(B) \tag{5.17}$$

is an $\mathscr{A}$-Markov–kernel, i.e., $\mathfrak{K}(x,\cdot)$ is a probability measure on $\mathscr{B}_1$ for each $x \in X$ and $\mathfrak{K}(\cdot,B)$ is $\mathscr{A}$-measurable for each $B \in \mathscr{B}$.

The first statement of the claim is trivial. The second one will be proved now. Consider the set $\mathscr{D} = \{C \in \mathscr{B}_1 \mid \mathfrak{K}(\cdot,C) \text{ is } \mathscr{A}\text{-measurable}\}$. This is a Dynkin system (see [Bauer 1972]) on $[0,1[$. Indeed, $[0,1[\, \in \mathscr{D}$ since for each $x \in X$

$$\mathfrak{K}(x,[0,1[\,) = \mathbf{q}_x([0,1[\,) = 1.$$

Also, if $C, D \in \mathscr{D}$ and $C \subseteq D$, then for each $x \in X$

$$\mathfrak{K}(x, D \setminus C) = \mathbf{q}_x(D) - \mathbf{q}_x(C) = \mathfrak{K}(x,D) - \mathfrak{K}(x,C),$$

and this shows that $\mathfrak{K}(x, D \setminus C)$ is $\mathscr{A}$-measurable too. Finally, if $(C_n)_{n\in\mathbb{N}}$ is a sequence of pairwise disjoint sets in $\mathscr{D}$, then we have for each $x \in X$

$$\mathfrak{K}(x, \bigcup_{n=1}^{\infty} C_n) = \mathbf{q}_x(x, \bigcup_{n=1}^{\infty} C_n) = \sum_{n=1}^{\infty} \mathbf{q}_x(C_n) = \sum_{n=1}^{\infty} \mathfrak{K}(x, C_n),$$

and, subsequently, $\mathfrak{K}(\cdot, \bigcup_{n=1}^{\infty} C_n)$ is also measurable. Thus, $\mathscr{D}$ is a Dynkin system on $[0,1[$. Now, observe that for all $\alpha,\beta \in [0,1]$ with $\alpha < \beta$ and for each $x \in X$ we have

$$\mathfrak{K}(x,[\alpha,\beta[\,) = \mathbf{q}_x([\alpha,\beta[\,) = h_x(\beta) - h_x(\alpha) = g_\beta(x) - g_\alpha(x),$$

and this implies that $\mathfrak{K}(\cdot,[\alpha,\beta[\,)$ is $\mathscr{A}$-measurable, i.e., $[\alpha,\beta[\, \in \mathscr{D}$. Thus the Dynkin system $\mathscr{D}$ contains all intervals $[\alpha,\beta[\, \subseteq [0,1[$ which means $\mathscr{D} = \mathscr{B}_1$, and so Claim 2 is proved.

Now we have to show that the $\mathscr{A}$-Markov–kernel $\mathfrak{K}$ defined in (5.17) indeed satisfies (5.9). To this end, let A be an $\mathscr{A}$-measurable step function $A = \sum_{i=1}^{n} \alpha_i \cdot A_i$, where all $\alpha_i \in [0,1]$ and the sets $A_1, A_2, \ldots, A_n$ form a partition of X. Then

$$\begin{aligned}
\boldsymbol{m}(A) &= \sum_{i=1}^{n} \boldsymbol{m}(\alpha_i \cdot A_i) \\
&= \sum_{i=1}^{n} \int_{A_i} g_{\alpha_i} \, d\overset{\vee}{\boldsymbol{m}} \\
&= \sum_{i=1}^{n} \int_{A_i} \mathfrak{K}(x, [0, \alpha_i[) \, d\overset{\vee}{\boldsymbol{m}}(x) \\
&= \sum_{i=1}^{n} \int_{X} \mathfrak{K}(x, [0, \alpha_i[) \cdot A_i(x) \, d\overset{\vee}{\boldsymbol{m}}(x) \\
&= \int_{X} \mathfrak{K}(x, [0, A(x)[) \, d\overset{\vee}{\boldsymbol{m}}(x).
\end{aligned}$$

If A is an $\mathscr{A}$-measurable fuzzy set and $(A_n)_{n\in\mathbb{N}}$ is a nondecreasing sequence of $\mathscr{A}$-measurable step functions converging pointwise to A, then

$$\begin{aligned}
\boldsymbol{m}(A) &= \sup_{n\in\mathbb{N}} \boldsymbol{m}(A_n) \\
&= \sup_{n\in\mathbb{N}} \int_{X} \mathfrak{K}(x, [0, A_n(x)[) \, d\overset{\vee}{\boldsymbol{m}}(x) \\
&= \int_{X} \mathfrak{K}(x, [0, A(x)[) \, d\overset{\vee}{\boldsymbol{m}}(x),
\end{aligned}$$

i.e., (5.9) holds.

Finally, we have to show the uniqueness of $\mathfrak{K}$. Since for $\mathscr{A}$-measurable functions $u, v : X \longrightarrow \mathbb{R}$ we can conclude $u = v$ $\overset{\vee}{\boldsymbol{m}}$-a.e. if for all $M \in \mathscr{A}$ we have

$$\int_M u \, d\overset{\vee}{\boldsymbol{m}} = \int_M v \, d\overset{\vee}{\boldsymbol{m}},$$

it follows that $\mathfrak{K}$ is $\overset{\vee}{\boldsymbol{m}}$-a.e. unique in the sense that any other $\mathscr{A}$-Markov–kernel satisfying (5.9) may differ from $\mathfrak{K}$ on an $\overset{\vee}{\boldsymbol{m}}$-null subset of X only. □

5.9 Taking into account Theorem 2.7, Proposition 5.5 and Remark 5.6(ii) we can restate Theorem 5.8 in the following form:

Corollary. *If T_s is a fundamental t-norm with $s \in [0,\infty]$, $\mathscr{T}$ is a generated T_s-tribe and $\boldsymbol{m}$ is a finite monotone T_s-measure on $\mathscr{T}$, then there exists a unique measure $\overset{\vee}{\boldsymbol{m}}$ on $\mathscr{T}^{\vee}$, namely, the restriction of $\boldsymbol{m}$ to $\mathscr{T}^{\vee}$, and an $\overset{\vee}{\boldsymbol{m}}$-a.e. uniquely* *determined $\mathscr{T}^{\vee}$-Markov–kernel $\mathfrak{K} : X \times \mathscr{B}_1 \longrightarrow \mathbb{R}$ such that the representation (5.9) holds for all $A \in \mathscr{T}$.*

6 Integral Representation of T_∞-Measures

6.1 The First Representation Theorem (Theorem 5.8) and Corollary 5.9 show that monotone finite measures based on fundamental t-norms T_s and defined on generated T_s-tribes can be represented as integrals of Markov-kernels. It is clear that this holds for $T = T_\infty$ too. However, in this particular case the result is true even if $\mathscr{T}$ is not generated. This is based on the following properties specific to T_∞-measures:

Proposition. *([Butnariu 1986]). If $\mathscr{T}$ is a T_∞-tribe and if $\boldsymbol{m}$ is a finite T_∞-measure on $\mathscr{T}$, then*

(i) $\boldsymbol{m}$ is monotone if and only if $\boldsymbol{m}$ is nonnegative;

(ii) if $\boldsymbol{m}$ is nonnegative, then $\boldsymbol{m}$ is continuous in the sense that

$$\lim_{n\to\infty} \boldsymbol{m}(A_n) = \boldsymbol{m}(A) \quad \textit{whenever} \quad (A_n)_{n\in\mathbb{N}} \subseteq \mathscr{T} \textit{ and } \lim_{n\to\infty} A_n = A. \tag{6.1}$$

PROOF: In (i) the "only if" part is trivial. Assume that $\boldsymbol{m}$ is nonnegative, $A, B \in \mathscr{T}$ and $A \leq B$. Then $B = A\,\mathbf{S}_\infty\,(B\,\mathbf{T}_\infty\,A^c)$ and $A\,\mathbf{T}_\infty\,(B\,\mathbf{T}_\infty\,A^c) = \emptyset$. Thus

$$\begin{aligned} \boldsymbol{m}(B) &= \boldsymbol{m}(A\,\mathbf{S}_\infty\,(B\,\mathbf{T}_\infty\,A^c)) + \boldsymbol{m}(A\,\mathbf{T}_\infty\,(B\,\mathbf{T}_\infty\,A^c)) \\ &= \boldsymbol{m}(A) + \boldsymbol{m}(B\,\mathbf{T}_\infty\,A^c), \end{aligned}$$

i.e., $\boldsymbol{m}(B) \geq \boldsymbol{m}(A)$.

In order to prove (ii), let $(A_n)_{n\in\mathbb{N}}$ be a nonincreasing sequence in $\mathscr{T}$ converging pointwise to A. Then

$$\lim_{n\to\infty} \boldsymbol{m}(A_n) = \boldsymbol{m}(X) - \lim_{n\to\infty} \boldsymbol{m}(A_n^c) = \boldsymbol{m}(X) - \boldsymbol{m}(A^c) = \boldsymbol{m}(A)$$

because of the continuity from the left of $\boldsymbol{m}$ and since for all $C \in \mathscr{T}, C$ and C^c are T_∞-disjoint and satisfy $C\,\mathbf{S}_\infty\,C^c = X$. Now, if $(A_n)_{n\in\mathbb{N}}$ is a (not necessarily monotone) sequence in $\mathscr{T}$ converging to A, then

$$\mathop{\mathbf{T}_0}_{k=1}^{\infty}(\mathop{\mathbf{S}_0}_{n=k}^{\infty} A_n) = A = \mathop{\mathbf{S}_0}_{k=1}^{\infty}(\mathop{\mathbf{T}_0}_{n=k}^{\infty} A_n),$$

and $\mathscr{T}$ is a T_0-tribe (cf. Theorem 2.7). Therefore, we have

$$\boldsymbol{m}(\mathop{\mathbf{T}_0}_{k=n}^{\infty} A_k) \le \boldsymbol{m}(A_n) \le \boldsymbol{m}(\mathop{\mathbf{S}_0}_{k=n}^{\infty} A_k)$$

and, taking here the limit $n \to \infty$ we obtain $\boldsymbol{m}(A) = \lim_{n\to\infty} \boldsymbol{m}(A_n)$. □

6.2 The following Representation Theorem of T_∞-measures is essentially Theorem 2.6(c) of BUTNARIU [1987b]. We present it here with an alternative proof.

Theorem. (Second Representation Theorem) *If $\mathscr{T}$ is a T_∞-tribe and $\boldsymbol{m}$ is a finite nonnegative T_∞-measure on it, then there exists a unique measure $\overset{\vee}{\boldsymbol{m}}$ on $\mathscr{T}^\vee$, namely, the restriction of $\boldsymbol{m}$ to $\mathscr{T}^\vee$, such that for each A in $\mathscr{T}$*

$$\boldsymbol{m}(A) = \int_X A(x)\, \mathrm{d}\overset{\vee}{\boldsymbol{m}}(x). \tag{6.2}$$

PROOF: It is clear that (6.2) implies that $\overset{\vee}{\boldsymbol{m}}$ has to be the restriction of $\boldsymbol{m}$ to $\mathscr{T}^\vee$.

Claim 1. If $A \in \mathscr{T}$, $\alpha, \beta \in [0,1]$ and $\alpha < \beta$, then the set

$$A_{\alpha,\beta} = \{x \in X \mid \alpha < A(x) \le \beta\} \tag{6.3}$$

belongs to $\mathscr{T}^\vee$, and the product $A \cdot A_{\alpha,\beta} = A\,\mathbf{T}_\infty\, A_{\alpha,\beta} \in \mathscr{T}$, and

$$\alpha \cdot \overset{\vee}{\boldsymbol{m}}(A_{\alpha,\beta}) \le \boldsymbol{m}(A \cdot A_{\alpha,\beta}). \tag{6.4}$$

The first assertion follows from Theorem 3.2. The second results from the fact that for each $M \in \mathscr{T}^\vee$ we have $A \cdot M = A\,\mathbf{T}_\infty\, M \in \mathscr{T}$. Now, in order to prove (6.4), it is sufficient to show that this inequality holds for all A in $\mathscr{T}$ and for all $\alpha, \beta \in [0,1]$, such that $\alpha < \beta$ and α has a finite binary representation

$$\alpha = \sum_{i=1}^{k} \frac{a_i}{2^i}$$

with $k \in \mathbb{N}$ and $a_i \in \{0,1\}$ for each $i \in \{1,\dots,k\}$. Suppose that (6.4) holds whenever α has a finite binary representation. For all $0 < \alpha < \beta \le 1$ we can find a sequence $(\alpha_n)_{n\in\mathbb{N}}$ which is nonnegative, nonincreasing and converges to α, and such that each α_n has a finite binary representation as above. Using the continuity of $\boldsymbol{m}$ (see Proposition 6.1(ii)) we get

$$\alpha \cdot \overset{\vee}{\boldsymbol{m}}(A_{\alpha,\beta}) = \lim_{n\to\infty} \alpha_n \cdot \overset{\vee}{\boldsymbol{m}}(A_{\alpha_n,\beta}) \le \lim_{n\to\infty} \boldsymbol{m}(A \cdot A_{\alpha_n,\beta}) = \boldsymbol{m}(A \cdot A_{\alpha,\beta}),$$

since $(A_{\alpha_n,\beta})_{n\in\mathbb{N}} \searrow A_{\alpha,\beta}$. Now assume that α has the form as above. If $\alpha = 0$ there is nothing to prove. Suppose $\alpha > 0$. In this situation we proceed by induction upon the number k in the binary representation of α. If $k = 1$, then $\alpha = 0$ or $\alpha = \frac{1}{2}$. In the first case (6.4) clearly holds. In the second case we have

$$A_{\alpha,\beta} = (A \cdot A_{\alpha,\beta})\, \mathbf{S}_\infty\, (A \cdot A_{\alpha,\beta}),$$

and this implies

$$\begin{aligned} \check{\boldsymbol{m}}\, (A_{\alpha,\beta}) &= \boldsymbol{m}(A_{\alpha,\beta}) \\ &= \boldsymbol{m}(A \cdot A_{\alpha,\beta}) + \boldsymbol{m}(A \cdot A_{\alpha,\beta}) - \boldsymbol{m}[(A \cdot A_{\alpha,\beta})\, \mathbf{T}_\infty\, (A \cdot A_{\alpha,\beta})] \\ &\leq 2 \cdot \boldsymbol{m}(A \cdot A_{\alpha,\beta}), \end{aligned}$$

which is exactly (6.4) with $\alpha = \frac{1}{2}$. Suppose that (6.4) holds for all $A \in \mathscr{T}$ and for all α, β in $[0,1]$ such that $\alpha < \beta$ and α has a binary representation with at most k binary digits. Put

$$\alpha = \sum_{i=1}^{k+1} \frac{a_i}{2^i}$$

with $\alpha_i \in \{0,1\}$ for each $i \in \{1, \ldots, m+1\}$ and suppose that $\alpha_{k+1} \neq 0$. Then $\alpha = \frac{\delta}{2}$ where

$$\delta = \sum_{i=1}^{k+1} \frac{a_i}{2^{i-1}} \tag{6.5}$$

Case 1.1. $\delta < 1$ and $\beta < \frac{1}{2}$. Then $A_{\alpha,\beta} = (A\, \mathbf{S}_\infty\, A)_{\delta,2\beta}$, and using the inductive assumption for the set $(A\, \mathbf{S}_\infty\, A)_{\delta,2\beta}$ (this is possible since the sum in (6.5) has at most k nonzero terms in our case) we get

$$\delta \cdot \check{\boldsymbol{m}}\, (A_{\alpha,\beta}) \leq \boldsymbol{m}[(A\, \mathbf{S}_\infty\, A) \cdot A_{\alpha,\beta}].$$

Observing that

$$(A\, \mathbf{S}_\infty\, A) \cdot A_{\alpha,\beta} = (A \cdot A_{\alpha,\beta})\, \mathbf{S}_\infty\, (A \cdot A_{\alpha,\beta}),$$

we obtain

$$\begin{aligned} \delta \cdot \check{\boldsymbol{m}}\, (A_{\alpha,\beta}) &\leq \boldsymbol{m}[(A\, \mathbf{S}_\infty\, A) \cdot A_{\alpha,\beta}] \\ &\leq \boldsymbol{m}[(A \cdot A_{\alpha,\beta})\, \mathbf{S}_\infty\, (A \cdot A_{\alpha,\beta})] \\ &= 2 \cdot \boldsymbol{m}(A \cdot A_{\alpha,\beta}) - \boldsymbol{m}[(A \cdot A_{\alpha,\beta})\, \mathbf{T}_\infty\, (A \cdot A_{\alpha,\beta})] \\ &\leq 2 \cdot \boldsymbol{m}(A \cdot A_{\alpha,\beta}) \end{aligned}$$

by the monotonicity and additivity of $\boldsymbol{m}$ (cf. Proposition 6.1(i)). This implies (6.4) in this case.

Case 1.2. $\delta < 1$ and $\beta = \frac{1}{2}$. Consider a sequence $(\gamma_n)_{n\in\mathbb{N}}$ in $]\alpha, \beta[$ which converges increasingly to β. The sets A_{α,γ_n} satisfy the conditions of Case 1.1. Thus we have for each $n \in \mathbb{N}$

$$\alpha \cdot \check{\boldsymbol{m}}\, (A_{\alpha,\gamma_n}) \leq \boldsymbol{m}(A \cdot A_{\alpha,\gamma_n}). \tag{6.6}$$

The continuity of $\boldsymbol{m}$ (cf. Proposition 6.1(ii)) and the convergence

$$(A_{\alpha,\gamma_n})_{n\in\mathbb{N}} \nearrow \mathop{\mathbf{S}_0}_{n=1}^{\infty} A_{\alpha,\gamma_n}$$

imply

$$\alpha \cdot \check{\boldsymbol{m}} \, (\mathop{\mathbf{S}_0}_{n=1}^{\infty} A_{\alpha,\gamma_n}) \leq \boldsymbol{m}[A \cdot (\mathop{\mathbf{S}_0}_{n=1}^{\infty} A_{\alpha,\gamma_n})], \tag{6.7}$$

where

$$(\bigcup_{n=1}^{\infty} A_{\alpha,\gamma_n}) \cup \{x \in X \mid A(x) = \beta\} = A_{\alpha,\beta}.$$

Since the sets in the latter union are disjoint, we may write

$$\begin{aligned} \alpha \cdot \check{\boldsymbol{m}} \, (A_{\alpha,\beta}) &= \alpha \cdot \check{\boldsymbol{m}} \, (\bigcup_{n=1}^{\infty} A_{\alpha,\gamma_n} + \alpha \cdot \check{\boldsymbol{m}} \, (\{A = \beta\}) \\ &\leq \boldsymbol{m}[A \cdot (\mathop{\mathbf{S}_0}_{n=1}^{\infty} A_{\alpha,\gamma_n})] + \alpha \cdot \check{\boldsymbol{m}} \, (\{A = \beta\}). \end{aligned} \tag{6.8}$$

Since for $\beta = \frac{1}{2}$ the fuzzy sets B_1 and B_2 defined by

$$B_1(x) = B_2(x) = \begin{cases} 0 & \text{if} \quad A(x) \neq \beta \\ \beta & \text{if} \quad A(x) = \beta \end{cases}$$

are T_∞-disjoint and are contained in $\mathscr{T}$, we have $\boldsymbol{m}(B_1 \, \mathbf{S}_\infty \, B_2) = 2 \cdot \boldsymbol{m}(\beta \cdot \{A = \beta\})$ (cf. Remark 5.2(i)) and $B_1 \, \mathbf{S}_\infty \, B_2 = \{A = \beta\}$ according to (1.22). Hence,

$$\boldsymbol{m}(\beta \cdot \{A = b\}) = \beta \cdot \check{\boldsymbol{m}} \, (\{A = \beta\}) \geq \alpha \cdot \boldsymbol{m}(\{A = \beta\}).$$

Combining this with (6.8) and using the T_∞-additivity of $\boldsymbol{m}$ we deduce

$$\begin{aligned} \alpha \cdot \boldsymbol{m}(A_{\alpha,\beta}) &\leq \boldsymbol{m}[A \cdot (\mathop{\mathbf{S}_0}_{n=1}^{\infty} A_{\alpha,\gamma_n})] + \boldsymbol{m}(\beta \cdot \{A = \beta\}) \\ &= \boldsymbol{m}[A \cdot (\{A = \beta\} \cup (\mathop{\mathbf{S}_0}_{n=1}^{\infty} A_{\alpha,\gamma_n}))] \\ &= \boldsymbol{m}(A \cdot A_{\alpha,\beta}). \end{aligned}$$

This proves (6.4) in this case.

Case 1.3. $\delta < 1$ and $\beta > \frac{1}{2}$. Then

$$A \cdot A_{\alpha,\beta} = (A \cdot A_{\alpha,\frac{1}{2}}) \, \mathbf{S}_\infty \, (A \cdot A_{\frac{1}{2},\beta})$$

where the fuzzy sets on the right hand side are T_∞-disjoint. Hence, according to Remark 5.2(i) we have

$$\boldsymbol{m}(A \cdot A_{\alpha,\beta}) = \boldsymbol{m}(A \cdot A_{\alpha,\frac{1}{2}}) + \boldsymbol{m}(A \cdot A_{\frac{1}{2},\beta}).$$

The first term on the right hand side satisfies the conditions of Case 1.2 and the second one meets the inductive assumption. Hence,

$$\boldsymbol{m}(A \cdot A_{\alpha,\beta}) \geq \alpha \cdot \check{\boldsymbol{m}}\,(A_{\frac{1}{2},\beta}) + \frac{1}{2} \cdot \check{\boldsymbol{m}}\,(A_{\frac{1}{2},\beta}) \geq \alpha \cdot [\check{\boldsymbol{m}}\,(A_{\alpha,\frac{1}{2}}) + \check{\boldsymbol{m}}\,(A_{\frac{1}{2},\beta})] = \alpha \cdot \check{\boldsymbol{m}}\,(A_{\alpha,\beta})$$

showing that (6.4) holds in this case too.

Case 1.4. $\delta \geq 1$. In this case α can be written as $\alpha = \frac{\varepsilon+1}{2}$ where

$$\varepsilon = \sum_{i=1}^{k} \frac{a_{i+1}}{2^i} \in [0,1].$$

We also have $A_{\alpha,\beta} = [A\,\mathbf{T}_\infty\, A]_{\varepsilon,2\rho}$ with $\rho = \beta - \frac{1}{2}$. Using the inductive assumption for the set $[A\,\mathbf{T}_\infty\, A]_{\varepsilon,2\rho}$ we obtain

$$\begin{aligned} \alpha \cdot \check{\boldsymbol{m}}\,(A_{\alpha,\beta}) &= \frac{1}{2} \cdot \check{\boldsymbol{m}}\,(A_{\alpha,\beta}) + \frac{\varepsilon}{2} \cdot \check{\boldsymbol{m}}\,(A_{\alpha,\beta}) \\ &= \frac{1}{2} \cdot \check{\boldsymbol{m}}\,(A_{\alpha,\beta}) + \frac{\varepsilon}{2} \cdot \check{\boldsymbol{m}}\left([A\,\mathbf{T}_\infty\, A]_{\varepsilon,2\rho}\right) \\ &\leq \frac{1}{2} \cdot \check{\boldsymbol{m}}\,(A_{\alpha,\beta}) + \frac{1}{2} \cdot \check{\boldsymbol{m}}\left([A\,\mathbf{T}_\infty\, A]_{\varepsilon,2\rho}\right) \\ &= \frac{1}{2} \cdot \check{\boldsymbol{m}}\,(A_{\alpha,\beta}) + \frac{1}{2}\boldsymbol{m}\left(A_{\alpha,\beta} \cdot [A\,\mathbf{T}_\infty\, A]_{\varepsilon,2\rho}\right). \end{aligned} \tag{6.9}$$

Observe that $A_{\alpha,\beta} \cdot (A\,\mathbf{T}_\infty\, A) = D\,\mathbf{T}_\infty\, C^c$ with $D = A_{\alpha,\beta}$ and $C = A_{\alpha,\beta} \cdot (A^c\,\mathbf{S}_\infty\, A^c)$. According to Remark 5.2(i), it follows that whenever $E, F \in \mathscr{T}$ and $E \geq F$ then

$$\boldsymbol{m}(E\,\mathbf{T}_\infty\, F^c) = \boldsymbol{m}(E) - \boldsymbol{m}(F). \tag{6.10}$$

Since we clearly have $D \geq C$, (6.9) combined with (6.10) gives

$$\alpha \cdot \check{\boldsymbol{m}}\,(A_{\alpha,\beta}) \leq \frac{1}{2} \cdot \check{\boldsymbol{m}}\,(A_{\alpha,\beta}) + \frac{1}{2} \cdot [\boldsymbol{m}(D) - \boldsymbol{m}(C)] = \check{\boldsymbol{m}}\,(A_{\alpha,\beta}) - \frac{1}{2} \cdot \boldsymbol{m}(C). \tag{6.11}$$

Now, taking into account $\delta \geq 1$, we deduce

$$C = (A^c \cdot A_{\alpha,\beta})\,\mathbf{S}_\infty\,(A^c \cdot A_{\alpha,\beta}) \quad \text{and} \quad (A^c \cdot A_{\alpha,\beta})\,\mathbf{T}_\infty\,(A^c \cdot A_{\alpha,\beta}) = \emptyset.$$

Thus $\boldsymbol{m}(C) = 2 \cdot \boldsymbol{m}(A^c \cdot A_{\alpha,\beta})$ by the T_∞-additivity of $\boldsymbol{m}$. Since

$$A^c \cdot A_{\alpha,\beta} = A_{\alpha,\beta}\,\mathbf{T}_\infty\,(A \cdot A_{\alpha,\beta})^c$$

we get

$$\boldsymbol{m}(C) = 2 \cdot [\boldsymbol{m}(A_{\alpha,\beta}) - \boldsymbol{m}(A \cdot A_{\alpha,\beta})]$$

because of (6.10). Substituting this into (6.11) we obtain (6.4), and Claim 1 is completely proved.

Claim 2. If $A \in \mathscr{T}$, $0 \leq \alpha < \beta \leq 1$ and

$$\bar{A}_{\alpha,\beta} = \{x \in X \mid \alpha < A(x) < \beta\} \tag{6.12}$$

then $\bar{A}_{\alpha,\beta} \in \mathscr{T}^\vee$, $A \cdot \bar{A}_{\alpha,\beta} \in \mathscr{T}$ and

$$\boldsymbol{m}(A \cdot \bar{A}_{\alpha,\beta}) \leq \beta \cdot \check{\boldsymbol{m}}\,(\bar{A}_{\alpha,\beta}). \tag{6.13}$$

The first two assertions are obvious. We only have to prove (6.13). To this end, observe that

$$A_{\alpha,\beta} = \overline{A^{\mathrm{c}}}_{\beta',\alpha'} \tag{6.14}$$

where $\alpha' = 1 - \alpha$ and $\beta' = 1 - \beta$. Then

$$\begin{aligned}
\boldsymbol{m}[(A \cdot \bar{A}_{\alpha,\beta})^{\mathrm{c}}] &= \check{\boldsymbol{m}}\,(X) - \boldsymbol{m}(A \cdot \bar{A}_{\alpha,\beta}) \\
&= \check{\boldsymbol{m}}\,(\bar{A}_{\alpha,\beta}) + \check{\boldsymbol{m}}\,([\bar{A}_{\alpha,\beta}]^{\mathrm{c}}) - \boldsymbol{m}(A \cdot \bar{A}_{\alpha,\beta}) \\
&= \boldsymbol{m}(A^{\mathrm{c}} \cdot \bar{A}_{\alpha,\beta}) + \check{\boldsymbol{m}}\,([\bar{A}_{\alpha,\beta}]^{\mathrm{c}}) \\
&= \boldsymbol{m}(A^{\mathrm{c}} \cdot A^{\mathrm{c}}_{\beta',\alpha'}) + \check{\boldsymbol{m}}\,([\bar{A}_{\alpha,\beta}]^{\mathrm{c}}) \\
&\geq \beta' \cdot \check{\boldsymbol{m}}\,(X) + \check{\boldsymbol{m}}\,([\bar{A}_{\alpha,\beta}]^{\mathrm{c}}) \\
&= \check{\boldsymbol{m}}\,(X) - \beta \cdot \check{\boldsymbol{m}}\,(\bar{A}_{\alpha,\beta}),
\end{aligned}$$

where the forth equality is a consequence of (6.14), and the inequality follows from Claim 1. This implies (6.13), and Claim 2 is proved.

In order to complete the proof, choose $A \in \mathscr{T}$. For a given $n \in \mathbb{N}$ denote

$$G_{n,i} = \begin{cases} \{x \in X \mid A(x) = 0\} & \text{if } i = 0, \\ A_{(i-1)/2^n, i/2^n} & \text{if } i \in \{1, 2, \ldots, 2^n\}, \end{cases}$$

$$H_{n,i} = \begin{cases} \bar{A}_{i/2^n, (i+1)/2^n} & \text{if } i \in \{0, 1, \ldots, 2^n - 1\}, \\ \{x \in X \mid A(x) = 1\} & \text{if } i = 2^n. \end{cases}$$

From Claims 1 and 2 it follows that the step functions

$$s_n = \sum_{i=1}^{2^n} \frac{i-1}{2^n} \cdot G_{n,i},$$

$$t_n = \sum_{i=1}^{2^n} \frac{i}{2^n} \cdot H_{n,i},$$

are all $\mathcal{T}^{\vee}$-measurable. It is clear that $(s_n) \nearrow A$ and $(t_n) \searrow A$. Taking into account (6.4) and (6.13) we deduce

$$\int_X s_n \, d\overset{\vee}{m} = \sum_{i=1}^{2^n} \frac{i-1}{2^n} \cdot \overset{\vee}{m}(G_{n,i}) \leq \sum_{i=0}^{2^n} m(A \cdot G_{n,i}) = m(A)$$

and

$$\int_X t_n \, d\overset{\vee}{m} = \sum_{i=1}^{2^n} \frac{i}{2^n} \cdot \overset{\vee}{m}(H_{n,i}) \geq \sum_{i=0}^{2^n} m(A \cdot H_{n,i}) = m(A).$$

Taking the limit $n \to \infty$ in the last two inequalities we obtain (6.2), completing the proof of the Theorem. □

6.3 Remark. Comparing Theorem 6.2 with Corollary 5.9 it is easily seen that the T_∞-measures on a generated T_s-tribe with T_s a fundamental t-norm are exactly those T_s-measures for which the corresponding Markov–kernel $\mathfrak{K}$, in the representation (5.9), for each $x \in X$ can be written in the form $\mathfrak{K}(x, [\alpha, \beta[) = \beta - \alpha$. This suggests that the concept of Lebesgue integral can be axiomatically defined as a T_∞-measure.

7 Integral Representation of Monotone T-Measures on Generated Tribes

7.1 Theorem 6.2 not only drops the condition that $\mathcal{T}$ be generated in Corollary 5.9 in the case $T = T_\infty$, but also specifies the form of the Markov–kernel occuring in the representation (5.9) in this particular case (see Remark 6.3). For fundamental t-norms T_s with $s \in]0, \infty[$ and generated T_s-tribes it is also possible to specify the form of the Markov-kernel in (5.9). To be precise, we get the following result of (KLEMENT [1982]):

Theorem. (Third Representation Theorem). *If T_s is a fundamental t-norm with $s \in]0, \infty[$, $\mathcal{T}$ is a generated T_s-tribe and m is a monotone finite T_s-measure on $\mathcal{T}$, then there exists a unique measure $\overset{\vee}{m}$ on $\mathcal{T}^{\vee}$, namely, the restriction of m to $\mathcal{T}^{\vee}$, and an $\overset{\vee}{m}$-a. e. uniquely determined $\mathcal{T}^{\vee}$-measurable function $f : X \longrightarrow [0,1]$ such that for all $A \in \mathcal{T}$*

$$m(A) = \int_{\{A>0\}} [f + (1-f) \cdot A] \, d\overset{\vee}{m} \, . \tag{7.1}$$

PROOF: First, observe that if (7.1) holds for some measure $\overset{\vee}{m}$ on $\mathcal{T}^{\vee}$ and for some $\mathcal{T}^{\vee}$-measurable function f on X, then, necessarily, $\overset{\vee}{m}$ has to be the restriction of $\boldsymbol{m}$ to $\mathcal{T}^{\vee}$. Now let us show that the required function f exists. To this end, consider the $\mathcal{T}^{\vee}$-Markov–kernel $\mathfrak{K}$ corresponding to $\boldsymbol{m}$ in the representation (5.9). For each a in $[0,1]$ define the function $f_a : X \longrightarrow [0,1]$ by

$$f_a(x) = \mathfrak{K}(x, [0, a[\,). \tag{7.2}$$

Each f_a is $\mathcal{T}^{\vee}$-measurable since $\mathfrak{K}$ is a $\mathcal{T}^{\vee}$-Markov–kernel. We claim that there exists a function $f : X \longrightarrow [0,1]$ such that for all $a, b \in]0,1[$ with $a + b \leq 1$,

$$f_{a+b} = f_a + f_b - f \qquad \overset{\vee}{\boldsymbol{m}}\text{-a. e.} \tag{7.3}$$

In order to prove that, consider the sequences $(a_n)_{n\in\mathbb{N}}$ and $(b_n)_{n\in\mathbb{N}}$ defined as follows:

$$\begin{aligned} a_1 &= a & \qquad b_1 &= b \\ a_{n+1} &= S_s(a_n, b_n) & \qquad b_n &= T_s(a_n, b_n). \end{aligned}$$

Since $\mathcal{T}$ is generated, these sequences belong to $\mathcal{T}$ (cf. Theorem 3.3). Exactly as in the proof of Theorem 2.7 we show that the sequence $(a_n)_{n\in\mathbb{N}}$ is nondecreasing, $(b_n)_{n\in\mathbb{N}}$ is nonincreasing and converges to zero, and for each $n \in \mathbb{N}$ we have $a_n + b_n = a + b$. Define the $\mathcal{T}^{\vee}$-measurable function

$$f = \inf_{n\in\mathbb{N}} f_{b_n}. \tag{7.4}$$

For each M in $\mathcal{T}^{\vee}$, the fuzzy sets $A_n = a_n \cdot M$ and $B_n = b_n \cdot M$ are contained in $\mathcal{T}$ (cf. Theorem 2.7), and $(A_n)_{n\in\mathbb{N}} \nearrow A_1 \mathbf{S}_\infty B_1$ and $(B_n)_{n\in\mathbb{N}} \searrow \emptyset$. Also, for each $n \in \mathbb{N}$, we get $A_n + B_n = (a + b) \cdot M$. Induction upon n yields that for all $n \in \mathbb{N}$

$$\boldsymbol{m}(A_n) + \boldsymbol{m}(B_n) = \boldsymbol{m}(a \cdot M) + \boldsymbol{m}(b \cdot M).$$

Consequently, we have

$$\begin{aligned} \int_M f_{a+b}\, d\overset{\vee}{\boldsymbol{m}} &= \boldsymbol{m}[(a+b)\cdot M] \\ &= \boldsymbol{m}(\sup_{n\in\mathbb{N}} A_n) \\ &= \sup_{n\in\mathbb{N}} \boldsymbol{m}(A_n) \\ &= \boldsymbol{m}(a\cdot M) + \boldsymbol{m}(b \cdot M) - \inf_{n\in\mathbb{N}} \boldsymbol{m}(B_n) \\ &= \int_M (f_a + f_b)\, d\overset{\vee}{\boldsymbol{m}} - \inf_{n\in\mathbb{N}} \boldsymbol{m}(B_n) \end{aligned}$$

by (7.2), (5.9) and the continuity from the left of $\boldsymbol{m}$. From (7.2), (5.9) and the Monotone Convergence Theorem we also have

$$\inf_{n\in\mathbb{N}} \boldsymbol{m}(B_n) = \inf_{n\in\mathbb{N}} \int_M f_{b_n}\, d\overset{\vee}{\boldsymbol{m}} = \int_M \inf_{n\in\mathbb{N}} f_{b_n}\, d\overset{\vee}{\boldsymbol{m}} = \int_M f\, d\overset{\vee}{\boldsymbol{m}}\,.$$

Hence,

$$\int_M f_{a+b}\,\mathrm{d}\check{m} = \int_M (f_a + f_b - f)\,\mathrm{d}\check{m}$$

for each M in $\mathcal{T}^\vee$. This implies (7.3), and our claim is proved. Equation (7.3) is a Gaussian functional equation (see [ACZEL 1969]) with the border conditions $f_1 = 1$ and $f_0 = f$. The unique solution of such an equation has to be

$$f_a = f + (1-f)\cdot a \qquad \check{m}\text{ -a.e.}$$

for all a in $]0,1[$. Taking into account (7.2) this gives

$$\mathfrak{K}(x,[0,a[) = f + (1-f)\cdot a \qquad \check{m}\text{ -a. e.}$$

for $a \in [0,1]$. Inserting this into (5.9), we get (7.1) and the proof is complete. □

7.2 Theorem 7.1 is proven for T_s-measures on generated T_s-tribes when $s \in]0,\infty[$. Combining this result with Theorem 6.2 we get:

Corollary. *If T_s is a fundamental t-norm with $s \in]0,\infty]$, $\mathcal{T}$ a generated T_s-tribe and m a monotone finite T_s-measure on $\mathcal{T}$, then there exists a unique finite measure $\check{m}$ on $\mathcal{T}^\vee$, namely, the restriction of m to $\mathcal{T}^\vee$, and an $\check{m}$-a.e. uniquely determined $\mathcal{T}^\vee$-measurable function $f : X \longrightarrow [0,1]$, such that for all $A \in \mathcal{T}$ the equality (7.1) holds.*

7.3 It is obvious (cf. Theorems 2.7 and 6.2) that T_∞-measures on T_s-tribes with $s \in]0,\infty[$ are also T_s-measures (even if the T_s-tribe is not generated). However, not each T_s-measure is necessarily a T_∞-measure, even on generated T_s-tribes. The following shows that a T_s-measure on a generated T_s-tribe is a T_∞-measure (and hence an integral) only if it is *continuous from the right* at $\emptyset$ (and then, clearly, at each $A \in \mathcal{T}$).

Corollary. *Let $s \in]0,\infty[$, T_s be the corresponding fundamental t-norm and $\mathcal{T}$ be a T_s-tribe. A monotone finite T_s-measure m on $\mathcal{T}$ is a T_∞-measure on $\mathcal{T}$ if and only if it satisfies the following condition:*

$$\lim_{n\to\infty} m(A_n) = 0 \quad \textit{whenever} \quad (A_n)_{n\in\mathbb{N}} \subseteq \mathcal{T} \quad \textit{and} \quad (A_n)_{n\in\mathbb{N}} \searrow \emptyset. \tag{7.5}$$

PROOF: From Proposition 6.1 we already know that T_∞-measures are continuous and, therefore, satisfy (7.5). Conversely, assume that m is a T_s-measure on the generated T_s-tribe $\mathcal{T}$ satisfying (7.5). Then m can be represented in the form (7.1) and for each M in $\mathcal{T}^\vee$ the sequence $(M_n)_{n\in\mathbb{N}} = \left(\frac{1}{n}\cdot M\right)_{n\in\mathbb{N}}$ is contained in $\mathcal{T}$. Using the Monotone Convergence Theorem we deduce that for each $M \in \mathcal{T}^\vee$

$$0 = \lim_{n\to\infty} \int_M [f + (1-f)\cdot\frac{1}{n}]\,\mathrm{d}\check{m} = \int_M f\,\mathrm{d}\check{m}\,.$$

This implies that $f = 0$ $\check{\boldsymbol{m}}$-a. e., and rewriting (7.1) in this case we obtain for $A \in \mathscr{T}$

$$\boldsymbol{m}(A) = \int_{\{A>0\}} A \,\mathrm{d}\check{\boldsymbol{m}} = \int_X A \,\mathrm{d}\check{\boldsymbol{m}}$$

which immediately implies that $\boldsymbol{m}$ is a T_∞-measure. □

8 Decomposition of Monotone T-Measures

8.1 Theorem 7.1 shows that for $s \in]0,\infty[$ finite monotone T_s-measures $\boldsymbol{m}$ which are defined on *generated* tribes $\mathscr{T}$ differ from T_∞-measures (i.e., from integrals according to Theorem 6.2) only by functions of the form

$$A \longmapsto \int_{\{A>0\}} f \,\mathrm{d}\check{\boldsymbol{m}},$$

which are themselves monotone finite T_s-measures on $\mathscr{T}$. The question we deal with in this section is how much a T_s-measure defined on a *nongenerated* T_s-tribe differs from a T_∞-measure (i.e., from an integral). The results are drawn from [Butnariu & Klement 1991]. However, the proofs there are somewhat incomplete, and we present them here in completed form, incorporating some suggestions due to R. Mesiar [personal communication].

8.2 Proposition. *Let $s \in]0,\infty]$, $\mathscr{T}$ be a T_s-tribe and $\boldsymbol{m} : \mathscr{T} \longrightarrow \mathbb{R}$ a finite monotone T_s-measure on $\mathscr{T}$. Then there exists a unique pair $(\boldsymbol{m}_\infty, \boldsymbol{m}_s)$ of functions from $\mathscr{T}$ to $[0,\infty[$ such that:*

(i) $\boldsymbol{m}_\infty$ is a T_∞-measure on $\mathscr{T}$;

(ii) $\boldsymbol{m}_s$ is a T_s-measure on $\mathscr{T}$;

(iii) $\boldsymbol{m} = \boldsymbol{m}_\infty + \boldsymbol{m}_s$;

(iv) $\boldsymbol{m}_\infty$ is the greatest T_∞-measure in the sense that we have $\boldsymbol{m}_\infty \geq \boldsymbol{n}$ whenever $\boldsymbol{n} : \mathscr{T} \longrightarrow [0,\infty[$ is a T_∞-measure such that $\boldsymbol{m} - \boldsymbol{n}$ is monotone;

(v) there exists a unique measure $\check{\boldsymbol{m}}$ on $\mathscr{T}^\vee$, namely, the restriction of $\boldsymbol{m}$ to $\mathscr{T}^\vee$, and an $\check{\boldsymbol{m}}$-a e. unique $\mathscr{T}^\vee$-measurable function $f : X \longrightarrow [0,1]$ such that for all $A \in \mathscr{T}$

$$\boldsymbol{m}_\infty(A) = \int_X (1-f) A \,\mathrm{d}\check{\boldsymbol{m}} \tag{8.1}$$

and for all $M \in \mathcal{T}^\vee$

$$\boldsymbol{m}_s(M) = \int_M f \, \mathrm{d}\overset{\vee}{\boldsymbol{m}} \,. \tag{8.2}$$

PROOF: Let $\mathcal{M}(\boldsymbol{m})$ be the set of all T_∞-measures $\boldsymbol{p} : \mathcal{T} \longrightarrow [0, \infty[$ such that $\boldsymbol{m} - \boldsymbol{p}$ is monotone (and, therefore, a T_s-measure due to Remark 5.6(ii) and 7.3). The set $\mathcal{M}(\boldsymbol{m})$ is nonempty since it contains the zero measure. $\mathcal{M}(\boldsymbol{m})$ is naturally equipped with the partial order

$$\boldsymbol{p} \leq \boldsymbol{q} \quad \text{if and only if} \quad \boldsymbol{p}(A) \leq \boldsymbol{q}(A) \quad \text{for all} \quad A \in \mathcal{T}. \tag{8.3}$$

Claim 1. There exists a maximal element in $\mathcal{M}(\boldsymbol{m})$.

If $(\boldsymbol{p}_j)_{j \in J}$ is a chain in $\mathcal{M}(\boldsymbol{m})$, then the function $\boldsymbol{p} : \mathcal{T} \longrightarrow [0, \infty]$ defined by

$$\boldsymbol{p}(A) = \sup_{j \in J} \boldsymbol{p}_j(A) \tag{8.4}$$

is a T_∞-valuation. Indeed, $\boldsymbol{p}(\emptyset) = 0$ clearly holds and for all $A, B \in \mathcal{T}$ we have

$$\begin{aligned} \boldsymbol{p}(A \, \mathbf{S}_\infty \, B) + \boldsymbol{p}(A \, \mathbf{T}_\infty \, B) &= \sup_{j \in J} \boldsymbol{p}_j(A \, \mathbf{S}_\infty \, B) + \sup_{j \in J} \boldsymbol{p}_j(A \, \mathbf{T}_\infty \, B) \\ &= \sup_{j \in J} [\boldsymbol{p}_j(A \, \mathbf{S}_\infty \, B) + \boldsymbol{p}_j(A \, \mathbf{T}_\infty \, B)] \\ &= \sup_{j \in J} [\boldsymbol{p}_j(A) + \boldsymbol{p}_j(B)] \\ &= \boldsymbol{p}(A) + \boldsymbol{p}(B) \end{aligned}$$

where the second and the fourth equality follow from the monotonicity of the chain $(\boldsymbol{p}_j(C))_{j \in J}$ for each $C \in \mathcal{T}$. It is clear that $\boldsymbol{p}$ is monotone. Hence, $0 \leq \boldsymbol{p}(A) \leq \boldsymbol{m}(A) \leq \boldsymbol{m}(X)$ for all $A \in \mathcal{T}$, implying that $\boldsymbol{p}$ is also finite. If $(A_n)_{n \in \mathbb{N}}$ is a nondecreasing sequence in $\mathcal{T}$, then

$$\begin{aligned} \lim_{n \to \infty} \boldsymbol{p}(A_n) &= \sup_{n \in \mathbb{N}} \boldsymbol{p}(A_n) \\ &= \sup_{n \in \mathbb{N}} (\sup_{j \in J} \boldsymbol{p}_j(A_n)) \\ &= \sup_{j \in J} (\sup_{n \in \mathbb{N}} \boldsymbol{p}_j(A_n)) \\ &= \sup_{j \in J} (\lim_{n \to \infty} \boldsymbol{p}_j(A_n)) \\ &= \boldsymbol{p}(\lim_{n \to \infty} A_n), \end{aligned}$$

showing that $\boldsymbol{p}$ is continuous from the left. Hence, $\boldsymbol{p}$ is a finite nonnegative T_∞-measure on $\mathcal{T}$. Moreover, $\boldsymbol{m} - \boldsymbol{p}$ is also monotone since for all $A, B \in \mathcal{T}$ with

$A \leq B$ it follows from $p_j \in \mathcal{M}(\boldsymbol{m})$ that

$$(\boldsymbol{m} - \boldsymbol{p})(A) = \inf_{j \in J}[(\boldsymbol{m} - \boldsymbol{p}_j)(A)] \leq \inf_{j \in J}[(\boldsymbol{m} - \boldsymbol{p}_j)(B)] = (\boldsymbol{m} - \boldsymbol{p})(B),$$

thus showing that $\boldsymbol{p} \in \mathcal{M}(\boldsymbol{m})$. In other words, each chain in $\mathcal{M}(\boldsymbol{m})$ has an upper bound in $\mathcal{M}(\boldsymbol{m})$ and, by Zorn's Lemma, $\mathcal{M}(\boldsymbol{m})$ contains a maximal element denoted $\boldsymbol{m}_\infty$. The T_∞-measure $\boldsymbol{m}_\infty$ is also a T_s-measure, and so is the difference

$$\boldsymbol{m}_s = \boldsymbol{m} - \boldsymbol{m}_\infty. \tag{8.5}$$

Because of $\boldsymbol{m}_\infty \in \mathcal{M}(\boldsymbol{m})$ we know that $\boldsymbol{m}_s$ is monotone and finite, and the pair $(\boldsymbol{m}_\infty, \boldsymbol{m}_s)$ satisfies properties (i) — (iii). The following claim, whose proof is due to R. Mesiar [personal communication], shows that $\mathcal{M}(\boldsymbol{m})$ is a directed set, thus immediately implying that $\boldsymbol{m}_\infty$ (and subsequently $\boldsymbol{m}_s$) are uniquely determined.

Claim 2. For all $\boldsymbol{q}_1, \boldsymbol{q}_2 \in \mathcal{M}(\boldsymbol{m})$ there exists a $\boldsymbol{q} \in \mathcal{M}(\boldsymbol{m})$ such that $\boldsymbol{q}_1 \leq \boldsymbol{q}$ and $\boldsymbol{q}_2 \leq \boldsymbol{q}$.

Since $\boldsymbol{q}_1, \boldsymbol{q}_2$ are T_∞-measures, we may write for $i \in \{1,2\}$

$$\boldsymbol{q}_i(A) = \int_X A \, \mathrm{d}\check{\boldsymbol{q}}_i,$$

where the finite σ-additive measure $\check{\boldsymbol{q}}_i$ is the restriction of $\boldsymbol{q}_i$ to $\mathcal{T}^\vee$. Then $\boldsymbol{q}_0 = \boldsymbol{q}_1 + \boldsymbol{q}_2$ is also a finite–additive measure on $\mathcal{T}^\vee$, and $\boldsymbol{q}_i$ is absolutely continuous with respect to $\boldsymbol{q}$ for $i \in \{1,2\}$. Then the Radon–Nikodym Theorem provides $\mathcal{T}^\vee$-measurable functions $h_1, h_2 : X \longrightarrow \mathbb{R}$ such that

$$\check{\boldsymbol{q}}_i\,(L) = \int_L h_i \, \mathrm{d}\boldsymbol{q}_0$$

for all $L \in \mathcal{T}^\vee$ and $i \in \{1,2\}$. Define the $\mathcal{T}^\vee$-measurable function $h : X \longrightarrow \mathbb{R}$ by $h = \max(h_1, h_2)$ and the finite σ-additive measure $\check{\boldsymbol{q}}$ on $\mathcal{T}^\vee$ by

$$\check{\boldsymbol{q}}\,(L) = \int_L h \, \mathrm{d}\boldsymbol{q}_0.$$

Then the function $\boldsymbol{q} : \mathcal{T} \longrightarrow [0, \infty[$ given by

$$\boldsymbol{q}(A) = \int_X A \, \mathrm{d}\check{\boldsymbol{q}}$$

is a finite monotone T_∞-measure such that $\boldsymbol{q}_i \leq \boldsymbol{q}$ for $i \in \{1,2\}$. The monotonicity of $\boldsymbol{m} - \boldsymbol{q}$ can be seen as follows. Consider the crisp sets $P = \{h = h_1\}$ and $Q = \{h > h_1\}$. Obviously $P, Q \in \mathcal{T}^\vee$ and $P \cup Q = X$ and $P \cap Q = \emptyset$. Note that for any fuzzy set $C \in [0,1]^X$ and for any crisp subset $L \subseteq X$ we always have $C \,\mathbf{T}_s\, L = C \cdot L$. Therefore, using the fact that for fuzzy sets $D, E \in \mathcal{T}$ with $D \leq P$ and $E \leq Q$ we always have $\boldsymbol{q}(D) = \boldsymbol{q}_1(D)$ and $\boldsymbol{q}(E) = \boldsymbol{q}_2(E)$, we obtain for all $A, B \in \mathcal{T}$ with $A \leq B$

$$\begin{aligned}
(\boldsymbol{m} - \boldsymbol{q})(A) &= (\boldsymbol{m} - \boldsymbol{q})(A\,\mathbf{T}_s\,P) + (\boldsymbol{m} - \boldsymbol{q})(A\,\mathbf{T}_s\,Q) \\
&= (\boldsymbol{m} - \boldsymbol{q}_1)(A\,\mathbf{T}_s\,P) + (\boldsymbol{m} - \boldsymbol{q}_2)(A\,\mathbf{T}_s\,Q) \\
&\leq (\boldsymbol{m} - \boldsymbol{q}_1)(B\,\mathbf{T}_s\,P) + (\boldsymbol{m} - \boldsymbol{q}_2)(B\,\mathbf{T}_s\,Q) \\
&= (\boldsymbol{m} - \boldsymbol{q})(B\,\mathbf{T}_s\,P) + (\boldsymbol{m} - \boldsymbol{q})(B\,\mathbf{T}_s\,Q) \\
&= (\boldsymbol{m} - \boldsymbol{q})(B),
\end{aligned}$$

showing that $\boldsymbol{m} - \boldsymbol{q}$ is monotone. This completes the proof of Claim 2 which together with Claim 1 implies that $(\boldsymbol{m}_\infty, \boldsymbol{m}_s)$ satisfies (iv) too.

In order to prove (v), we take $\overset{\vee}{\boldsymbol{m}}$ for the restriction of $\boldsymbol{m}$ to $\mathcal{T}^\vee$. We have to find a $\mathcal{T}^\vee$-measurable function $f : X \longrightarrow [0,1]$ such that (8.1) and (8.2) hold. To this end, observe that $\mathcal{T}$ is a T_∞-tribe by Theorem 2.7. Hence, according to Theorem 6.2, the T_∞-measure $\boldsymbol{m}_\infty$ can be written as

$$\boldsymbol{m}_\infty(A) = \int_X A \, \mathrm{d}\overset{\vee}{\boldsymbol{m}}_\infty \tag{8.6}$$

for each $A \in \mathcal{T}$ where $\overset{\vee}{\boldsymbol{m}}_\infty$ is the restriction of $\boldsymbol{m}_\infty$ to $\mathcal{T}^\vee$. From $\boldsymbol{m}_\infty \leq \boldsymbol{m}$ it follows that $\overset{\vee}{\boldsymbol{m}}_\infty$ is absolutely continuous with respect to $\overset{\vee}{\boldsymbol{m}}$. Because of $\overset{\vee}{\boldsymbol{m}}_\infty \leq \boldsymbol{m}$ there is a Radon–Nikodym derivative

$$\frac{\mathrm{d}(\tilde{\boldsymbol{m}})^\vee}{\mathrm{d}\overset{\vee}{\boldsymbol{m}}}$$

which maps X into $[0,1]$. Putting $f = 1 - \mathrm{d}\overset{\vee}{\boldsymbol{m}}_\infty / \mathrm{d}\overset{\vee}{\boldsymbol{m}}$, then f is also a $\mathcal{T}^\vee$-measurable function with values in $[0,1]$, and (8.6) implies (8.1). Now, taking into account (8.5), we can write for all crisp sets $L \in \mathcal{T}^\vee$

$$\boldsymbol{m}_s(L) = \boldsymbol{m}(L) = \overset{\vee}{\boldsymbol{m}}(L) - \overset{\vee}{\boldsymbol{m}}_\infty(L) = \overset{\vee}{\boldsymbol{m}}(L) - \int_L (1 - f)\, \mathrm{d}\overset{\vee}{\boldsymbol{m}} = \int_L f \, \mathrm{d}\overset{\vee}{\boldsymbol{m}},$$

and therefore, (8.2) also holds. Formulae (8.1), (8.2) and (8.5) imply that no other choice for $\overset{\vee}{\boldsymbol{m}}$ (and, subsequently, for f) is possible. □

8.3 Remark.

(i) The component $\boldsymbol{m}_s$ of a T_s-measure $\boldsymbol{m}$, which exists due to Proposition 8.2, is a *pure* T_s-measure in the sense that it has a zero T_∞-component: $(\boldsymbol{m}_s)_\infty = 0$. In fact, assuming the contrary contradicts the maximality of $\boldsymbol{m}_\infty$ in Proposition 8.2.

(ii) If the T_s-tribe $\mathcal{T}$ is generated, then (8.2) somehow holds for each $A \in \mathcal{T}$ (and not only for crisp sets in the σ-algebra $\mathcal{T}^{\vee}$). To be precise, we always have

$$\boldsymbol{m}_s(A) = \int_{\{A>0\}} f \, \mathrm{d}\overset{\vee}{\boldsymbol{m}} \, .$$

This follows from (7.1) in combination with (8.1) and (8.2), and the $\overset{\vee}{\boldsymbol{m}}$-a.e. uniqueness of f in (7.1).

8.4 Let $s \in]0, \infty[$ and $\mathcal{T}$ be a T_s-tribe. If $\boldsymbol{p}$ is a σ-additive measure on the σ-algebra $\mathcal{T}^{\vee}$ and if $g, h : X \longrightarrow [0, \infty]$ are two $\mathcal{T}^{\vee}$-measurable functions, then the function $\boldsymbol{m} : \mathcal{T} \longrightarrow [0, +\infty]$ defined by

$$\boldsymbol{m}(A) = \int_{\{A>0\}} (g + h \cdot A) \, \mathrm{d}\boldsymbol{p} \tag{8.7}$$

is a monotone T_s-measure on $\mathcal{T}$. Indeed, monotonicity is obvious as is $\boldsymbol{m}(\emptyset) = 0$; continuity from the left follows from the Monotone Convergence Theorem, taking into account that for a nondecreasing sequence $(A_n)_{n\in\mathbb{N}} \subseteq \mathcal{T}$ which converges pointwise to A we have

$$\bigcup_{n=1}^{n} \{A_n > 0\} = \{A > 0\}.$$

The T_s-additivity is shown as follows:

$$\begin{aligned}
&\boldsymbol{m}(A \, \mathbf{T}_s \, B) + \boldsymbol{m}(A \, \mathbf{S}_s \, B) \\
&= \int_{\{A>0\}\cap\{B>0\}} [g + h \cdot (A \, \mathbf{T}_s \, B)] \, \mathrm{d}\boldsymbol{p} + \int_{\{A>0\}\cup\{B>0\}} [g + h \cdot (A \, \mathbf{S}_s \, B)] \, \mathrm{d}\boldsymbol{p} \\
&= \int_{\{A>0\}\cap\{B>0\}} [2 \cdot g + h \cdot ((A \, \mathbf{T}_s \, B) + (A \, \mathbf{S}_s \, B))] \, \mathrm{d}\boldsymbol{p} \\
&\quad + \int_{\{A>0\}\cap\{B=0\}} [g + h(A \, \mathbf{S}_s \, B)] \, \mathrm{d}\boldsymbol{p} + \int_{\{A=0\}\cap\{B>0\}} [g + h(A \, \mathbf{S}_s \, B)] \, \mathrm{d}\boldsymbol{p} \\
&= \int_{\{A>0\}\cap\{B>0\}} [2 \cdot g + h \cdot (A + B)] \, \mathrm{d}\boldsymbol{p} + \int_{\{A>0\}\cap\{B=0\}} (g + h \cdot A) \, \mathrm{d}\boldsymbol{p} \\
&\quad + \int_{\{A=0\}\cap\{B>0\}} (g + h \cdot B) \, \mathrm{d}\boldsymbol{p} \\
&= \int_{\{A>0\}} (g + h \cdot A) \, \mathrm{d}\boldsymbol{p} + \int_{\{B>0\}} (g + h \cdot B) \, \mathrm{d}\boldsymbol{p} \\
&= \boldsymbol{m}(A) + \boldsymbol{m}(B).
\end{aligned}$$

Note that the measure $\boldsymbol{p}$ involved in the representation of $\boldsymbol{m}$ via (8.7) does not necessarily coincide with the restriction of $\boldsymbol{m}$ to $\mathscr{T}^\vee$.

Definition. A T_s-measure $\boldsymbol{m}$ on $\mathscr{T}$ which can be represented in the form (8.7) for some nonnegative σ-additive measure $\boldsymbol{p}$ on $\mathscr{T}^\vee$ and for some $\mathscr{T}^\vee$-measurable functions $g, h : X \longrightarrow [0, \infty]$ is said to be *generated (by* $\boldsymbol{p}, g$ *and* h*)*.

8.5 Theorem 7.1 shows that if $\mathscr{T}$ is generated, then all finite monotone T_s-measures on $\mathscr{T}$ are generated. From Theorem 6.2 we already know that T_∞-measures on $\mathscr{T}$ are always generated, even if $\mathscr{T}$ is not generated. Thus, it is natural to ask whether, in general, T_s-measures on T_s-tribes are automatically generated. In order to answer this question we introduce the following notion.

Definition. A T_s-measure $\boldsymbol{m}$ on a T_s-tribe $\mathscr{T}$ is called *monotonically irreducible* if it is monotone and if the zero measure on $\mathscr{T}$ is the only generated T_s-measure $\boldsymbol{q}$ on $\mathscr{T}$ such that $\boldsymbol{m} - \boldsymbol{q}$ is monotone.

It is obvious that a T_s-measure $\boldsymbol{m}$ is generated if and only if it can be extended to a T_s-measure on the generated T_s-tribe $\mathscr{T}^\circ$: if $\boldsymbol{m}$ is generated, then it is always defined on the whole generated tribe $\mathscr{T}^\circ$; the converse follows from Theorem 7.1. On the other hand, monotonically irreducible T_s-measures, with the exception of the zero measure, are not generated and, therefore, cannot be extended to $\mathscr{T}^\circ$.

8.6 Theorem. (Decomposition Theorem.) *Let $s \in]0, \infty]$, $\mathscr{T}$ be a T_s-tribe and $\boldsymbol{m}$ a finite monotone T_s-measure on $\mathscr{T}$. Then $\boldsymbol{m}$ can be uniquely decomposed into the sum of a monotonically irreducible and a generated T_s-measure, i.e., there exist a unique monotonically irreducible T_s-measure $\mathring{\boldsymbol{m}}$ on $\mathscr{T}$, a measure $\check{\boldsymbol{m}}$ on $\mathscr{T}^\vee$ (namely, the restriction of $\boldsymbol{m}$ to $\mathscr{T}^\vee$) and two $\check{\boldsymbol{m}}$-a.e. uniquely determined $\mathscr{T}^\vee$-measurable funtions $g, h : X \longrightarrow [0, 1]$, such that for all $A \in \mathscr{T}^\vee$*

$$\boldsymbol{m}(A) = \mathring{\boldsymbol{m}}(A) + \int_{\{A>0\}} (g + h \cdot A)\, \mathrm{d}\check{\boldsymbol{m}} . \tag{8.8}$$

PROOF: If $s = \infty$, then the result follows from Theorem 6.2 taking for g and h the constant functions $g(x) = 0$ and $h(x) = 1$, respectively, and for $\mathring{\boldsymbol{m}}$ the zero measure. Assume, therefore, that $s \in]0, \infty[$. In this case we proceed in several steps.

Claim 1. If $\boldsymbol{p}$ is a finite monotone T_s-measure on $\mathscr{T}$, then there exists a finite monotone T_s-measure $\tilde{\boldsymbol{p}}$ on the generated T_s-tribe $\mathscr{T}^\circ$ which is maximal in the sense that the difference $\tilde{\boldsymbol{p}} - \boldsymbol{q}$ is also monotone on $\mathscr{T}^\circ$ whenever $\boldsymbol{q}$ is a T_s-measure on $\mathscr{T}^\circ$ such that $\boldsymbol{p} - \boldsymbol{q}$ is monotone on $\mathscr{T}$.

In order to verify this, let $\mathscr{N}(\boldsymbol{p})$ be the set of all monotone T_s-measures $\boldsymbol{q}$ on $\mathscr{T}^\circ$ such that $\boldsymbol{p} - \boldsymbol{q}$ is monotone on $\mathscr{T}$. This set is partially ordered by the *dominance relation* defined by

$$\boldsymbol{q} \gg \boldsymbol{r} \quad \text{if and only if} \quad \boldsymbol{q} - \boldsymbol{r} \quad \text{is monotone on } \mathcal{T}^{\circ}. \tag{8.9}$$

Let $(\boldsymbol{q}_j)_{j \in J}$ be a chain in $\mathcal{N}(\boldsymbol{p})$, and for each $A \in \mathrm{T}^{\circ}$ define

$$\boldsymbol{q}(A) = \sup_{j \in J} \boldsymbol{q}_j(A).$$

Exactly as in the proof of Proposition 8.2, it follows that $\boldsymbol{q}$ is a monotone finite T_s-measure on $\mathcal{T}^{\circ}$. For $A, B \in \mathcal{T}$ with $A \leq B$ we have

$$\begin{aligned} \boldsymbol{p}(A) - \boldsymbol{q}(A) &= \inf_{j \in J} [\boldsymbol{p}(A) - \boldsymbol{q}_j(A)] \\ &\leq \inf_{j \in J} [\boldsymbol{p}(B) - \boldsymbol{q}_j(B)] \\ &= \boldsymbol{p}(B) - \boldsymbol{q}(B), \end{aligned}$$

i.e., $\boldsymbol{p} - \boldsymbol{q}$ is monotone on $\mathcal{T}$. For each $k \in J$ and for $A, B \in \mathcal{T}^{\circ}$ with $A \leq B$, the fact that $(\boldsymbol{q}_j)_{j \in J}$ is a chain induces

$$(\boldsymbol{q} - \boldsymbol{q}_k)(A) = \sup_{j \in J} [(\boldsymbol{q}_j - \boldsymbol{q}_k)(A)] \leq \sup_{j \in J} [(\boldsymbol{q}_j - \boldsymbol{q}_k)(B)] = (\boldsymbol{q} - \boldsymbol{q}_k)(B),$$

which means that $\boldsymbol{q} - \boldsymbol{q}_k$ is monotone for each $k \in J$, i.e., $\boldsymbol{q} \gg \boldsymbol{q}_k$. Hence, each chain in $\mathcal{N}(\boldsymbol{p})$ has an upper bound in $\mathcal{N}(\boldsymbol{p})$ and, by Zorn's Lemma, this implies that $\mathcal{N}(\boldsymbol{p})$, equipped with the dominance relation $\gg$, contains a maximal element, completing the proof of Claim 1.

The following claim shows that $\mathcal{N}(\boldsymbol{p})$ is also a directed set, implying that the maximal element is indeed the unique greatest element.

Claim 2. For all $\boldsymbol{q}_1, \boldsymbol{q}_2 \in \mathcal{N}(\boldsymbol{p})$ there exists a $\boldsymbol{q} \in \mathcal{N}(\boldsymbol{p})$ such that $\boldsymbol{q} \gg \boldsymbol{q}_1$ and $\boldsymbol{q} \gg \boldsymbol{q}_2$.

Let $\overset{\vee}{\boldsymbol{q}}_1$, $\overset{\vee}{\boldsymbol{q}}_2$ be the restrictions of $\boldsymbol{q}_1$ and $\boldsymbol{q}_2$ to $\mathcal{T}^{\vee}$, respectively, let (P, Q) be a Hahn–decomposition of the σ-additive signed measure $\overset{\vee}{\boldsymbol{q}}_1 - \overset{\vee}{\boldsymbol{q}}_2$, and define

$$\boldsymbol{q} : \mathcal{T}^{\circ} \longrightarrow [0, \infty]$$

by

$$\boldsymbol{q}(A) = \boldsymbol{q}_1(A\,\mathbf{T}_s\,P) + \boldsymbol{q}_2(A\,\mathbf{T}_s\,Q).$$

Obviously, $\boldsymbol{q}(\emptyset) = 0$ and $\boldsymbol{q}$ is continuous from the left. To see that $\boldsymbol{q}$ is also a T_s-valuation, take into account that P and Q are crisp subsets of X, and then we get for all $A, B \in \mathcal{T}^{\circ}$:

$$\begin{aligned}
&\boldsymbol{q}(A\,\mathbf{T}_s\,B)+\boldsymbol{q}(A\,\mathbf{S}_s\,B)=\\
&\quad= \boldsymbol{q}_1(A\,\mathbf{T}_s\,B\,\mathbf{T}_s\,P)+\boldsymbol{q}_2(A\,\mathbf{T}_s\,B\,\mathbf{T}_s\,Q)\\
&\qquad+\boldsymbol{q}_1((A\,\mathbf{S}_s\,B)\,\mathbf{T}_s\,P)+\boldsymbol{q}_2((A\,\mathbf{S}_s\,B)\,\mathbf{T}_s\,Q)\\
&\quad= \boldsymbol{q}_1((A\,\mathbf{T}_s\,P)\,\mathbf{T}_s\,(B\,\mathbf{T}_s\,P))+\boldsymbol{q}_1((A\,\mathbf{T}_s\,P)\,\mathbf{S}_s\,(B\,\mathbf{T}_s\,P))\\
&\qquad+\boldsymbol{q}_2((A\,\mathbf{T}_s\,Q)\,\mathbf{T}_s\,(B\,\mathbf{T}_s\,Q))+\boldsymbol{q}_2((A\,\mathbf{T}_s\,Q)\,\mathbf{S}_s\,(B\,\mathbf{T}_s\,Q))\\
&\quad= \boldsymbol{q}_1(A\,\mathbf{T}_s\,P)+\boldsymbol{q}_1(B\,\mathbf{T}_s\,P)+\boldsymbol{q}_2(A\,\mathbf{T}_s\,Q)+\boldsymbol{q}_2(B\,\mathbf{T}_s\,Q)\\
&\quad= \boldsymbol{q}(A)+\boldsymbol{q}(B).
\end{aligned}$$

The T_s-measure $\boldsymbol{q}$ is monotone since for fuzzy sets $A, B \in \mathcal{T}$ with $A \leq B$ we have

$$\boldsymbol{q}(A)=\boldsymbol{q}_1(A\,\mathbf{T}_s\,P)+\boldsymbol{q}_2(A\,\mathbf{T}_s\,Q)\leq \boldsymbol{q}_1(B\,\mathbf{T}_s\,P)+\boldsymbol{q}_2(B\,\mathbf{T}_s\,Q)=\boldsymbol{q}(B).$$

The monotone T_s-measure $\boldsymbol{q}$ is an element of $\mathcal{N}(\boldsymbol{q})$ since for fuzzy sets $A, B \in \mathcal{T}^\diamond$ with $A \leq B$ we get

$$\begin{aligned}
(\boldsymbol{p}-\boldsymbol{q})(A) &= \boldsymbol{p}(A)-\boldsymbol{q}(A)\\
&= \boldsymbol{p}(A\,\mathbf{T}_s\,P)+\boldsymbol{p}(A\,\mathbf{T}_s\,Q)-\boldsymbol{q}_1(A\,\mathbf{T}_s\,P)-\boldsymbol{q}_2(A\,\mathbf{T}_s\,Q)\\
&= (\boldsymbol{p}-\boldsymbol{q}_1)(A\,\mathbf{T}_s\,P)+(\boldsymbol{p}-\boldsymbol{q}_2)(A\,\mathbf{T}_s\,Q)\\
&\leq (\boldsymbol{p}-\boldsymbol{q}_1)(B\,\mathbf{T}_s\,P)+(\boldsymbol{p}-\boldsymbol{q}_2)(B\,\mathbf{T}_s\,Q)\\
&= \boldsymbol{p}(B)-\boldsymbol{q}(B)\\
&= (\boldsymbol{p}-\boldsymbol{q})(B),
\end{aligned}$$

where the inequality is a consequence of $\boldsymbol{q}_1, \boldsymbol{q}_2 \in \mathcal{N}(\boldsymbol{p})$. To complete the proof of Claim 2, it remains to show that $\boldsymbol{q} \gg \boldsymbol{q}_1$ and $\boldsymbol{q} \gg \boldsymbol{q}_2$, i.e., $\boldsymbol{q}-\boldsymbol{q}_1$ and $\boldsymbol{q}-\boldsymbol{q}_2$ are monotone on $\mathcal{T}^\diamond$. For each $A \in \mathcal{T}^\diamond$ we get

$$\begin{aligned}
(\boldsymbol{q}-\boldsymbol{q}_1)(A) &= \boldsymbol{q}_1(A\,\mathbf{T}_s\,P)+\boldsymbol{q}_2(A\,\mathbf{T}_s\,Q)-\boldsymbol{q}_1(A\,\mathbf{T}_s\,P)-\boldsymbol{q}_1(A\,\mathbf{T}_s\,Q)\\
&= (\boldsymbol{q}_2-\boldsymbol{q}_1)(A\,\mathbf{T}_s\,Q)\\
&\geq 0,
\end{aligned}$$

showing that $\boldsymbol{q}-\boldsymbol{q}_1$ is nonnegative. This also implies that the restriction of $(\boldsymbol{q}-\boldsymbol{q}_1)^\vee$ to the σ-algebra of all crisp sets of the form $L \cap Q$ with $L \in \mathcal{T}^\vee$ is nonnegative and, therefore, monotone. Applying Theorem 7.1 to the T_s-measure $\boldsymbol{q}_2-\boldsymbol{q}_1$ which is monotone on the T_s-tribe of all fuzzy subsets of Q contained in $\mathcal{T}^\diamond$, we get for all $A, B \in \mathcal{T}^\diamond$ with $A \leq B$ and for a suitable $\mathcal{T}^\vee$-measurable function $k : X \longrightarrow [0,1]$:

$$\begin{aligned}
(\boldsymbol{q}-\boldsymbol{q}_1)(A) &= (\boldsymbol{q}_2-\boldsymbol{q}_1)(A\,\mathbf{T}_s\,Q)\\
&= \int_{\{A\,\mathbf{T}_s\,Q>0\}} (k+(1-k)\cdot A)\,\mathrm{d}(\boldsymbol{q}_2-\boldsymbol{q}_1)^\vee
\end{aligned}$$

$$\begin{aligned}
&= \int_{\{A>0\}\cap Q} (k + (1-k)\cdot A)\,\mathrm{d}(\boldsymbol{q}_2 - \boldsymbol{q}_1)^{\vee} \\
&\le \int_{\{A>0\}\cap Q} (k + (1-k)\cdot B)\,\mathrm{d}(\boldsymbol{q}_2 - \boldsymbol{q}_1)^{\vee} \\
&\quad + \int_{\{A=0\}\cap\{B>0\}\cap Q} (k + (1-k)\cdot B)\,\mathrm{d}(\boldsymbol{q}_2 - \boldsymbol{q}_1)^{\vee} \\
&= \int_{\{B>0\}\cap Q} (k + (1-k)\cdot B)\,\mathrm{d}(\boldsymbol{q}_2 - \boldsymbol{q}_1)^{\vee} \\
&= (\boldsymbol{q}_2 - \boldsymbol{q}_1)(B\,\mathbf{T}_s\,Q) \\
&= (\boldsymbol{q} - \boldsymbol{q}_1)(B).
\end{aligned}$$

Monotonicity of $\boldsymbol{q} - \boldsymbol{q}_2$ is shown by complete analogy, interchanging the roles of $\boldsymbol{q}_1$ and $\boldsymbol{q}_2$ as well as of P and Q, thus completing the proof of Claim 2.

Now let $\tilde{\boldsymbol{m}}$ be the greatest T_s-measure (with respect to the dominance relation $\gg$) in $\mathcal{N}(\boldsymbol{m})$ resulting from Claims 1 and 2. Since $\tilde{\boldsymbol{m}}$ is defined on $\mathcal{T}^{\circ}$, Theorem 7.1 yields the representation

$$\tilde{\boldsymbol{m}}\,(A) = \int_{\{A>0\}} [l - (1-l)\cdot A]\,\mathrm{d}(\tilde{\boldsymbol{m}})^{\vee} \tag{8.10}$$

where $(\tilde{\boldsymbol{m}})^{\vee}$ is the restriction of $(\tilde{\boldsymbol{m}})$ to $(\mathcal{T}^{\circ})^{\vee} = \mathcal{T}^{\vee}$ and $l : X \longrightarrow [0,1]$ is a suitable $\mathcal{T}^{\vee}$-measurable function. Remark 8.3(ii) tells us that the unique decomposition $((\tilde{\boldsymbol{m}})_{\infty}, (\tilde{\boldsymbol{m}})_s)$ of $\tilde{\boldsymbol{m}}$ (cf. Proposition 8.2) is given by

$$(\tilde{\boldsymbol{m}})_{\infty}(A) = \int_X (1-l)\cdot A\,\mathrm{d}(\tilde{\boldsymbol{m}})^{\vee} \tag{8.11}$$

and

$$(\tilde{\boldsymbol{m}})_s(A) = \int_{\{A>0\}} (1-l)\,\mathrm{d}(\tilde{\boldsymbol{m}})^{\vee}. \tag{8.12}$$

Let $(\boldsymbol{m}_{\infty}, \boldsymbol{m}_s)$ be the unique decomposition of $\boldsymbol{m}$ resulting from Proposition 8.2, let $\overset{\vee}{\boldsymbol{m}}$ be the restriction of $\boldsymbol{m}$ to $\mathcal{T}^{\vee}$, and let $f : X \longrightarrow [0,1]$ be the $\mathcal{T}^{\vee}$-measurable function such that (8.1) and (8.2) hold.

Claim 3. $(\tilde{\boldsymbol{m}})_{\infty} = \boldsymbol{m}_{\infty}$.

From $\tilde{\boldsymbol{m}} \in \mathcal{N}(\boldsymbol{m})$ it follows that the function $\boldsymbol{r} : \mathcal{T} \longrightarrow \mathbb{R}$ defined by

$$\boldsymbol{r} = \boldsymbol{m} - \tilde{\boldsymbol{m}}$$

is a monotone T_s-measure. Using $\boldsymbol{m} - (\tilde{\boldsymbol{m}})_{\infty} = \boldsymbol{r} + (\tilde{\boldsymbol{m}})_s$, the T_s-measure $\boldsymbol{m} - (\tilde{\boldsymbol{m}})_{\infty}$ is monotone too. Since $\boldsymbol{m}_{\infty}$ satisfies Proposition 8.2(iv), we get

$$\boldsymbol{m}_{\infty} \ge (\tilde{\boldsymbol{m}})_{\infty} \tag{8.13}$$

Conversely, note that $\boldsymbol{m} - \boldsymbol{m}_\infty$ is a monotone T_s-measure on $\mathcal{T}$, which, together with the fact that $\tilde{\boldsymbol{m}}$ is the greatest element in $\mathcal{N}(\boldsymbol{m})$, implies that $\tilde{\boldsymbol{m}} \gg \boldsymbol{m}_\infty$, from which we deduce that $\tilde{\boldsymbol{m}} - \boldsymbol{m}_\infty$ is monotone, showing that $\boldsymbol{m}_\infty \in \mathcal{M}(\tilde{\boldsymbol{m}})$. Since $(\tilde{\boldsymbol{m}})_\infty$ is the greatest element of $\mathcal{M}(\tilde{\boldsymbol{m}})$, we get $(\tilde{\boldsymbol{m}})_\infty \geq \boldsymbol{m}_\infty$ and, together with (8.13), $(\tilde{\boldsymbol{m}})_\infty = \boldsymbol{m}_\infty$.

Claim 4. The monotone T_s-measure $\mathring{\boldsymbol{m}}\colon \mathcal{T} \longrightarrow [0, \infty[$ defined by

$$\mathring{\boldsymbol{m}} = \boldsymbol{m} - \tilde{\boldsymbol{m}} \tag{8.14}$$

is monotonically irreducible.

In the case $\tilde{\boldsymbol{m}} = \boldsymbol{m}$ the claim is proved by definition. Suppose that $\tilde{\boldsymbol{m}} \neq \boldsymbol{m}$, i.e., $\mathring{\boldsymbol{m}}$ is different from the zero measure. Then Claim 3 yields

$$\begin{aligned}
\mathring{\boldsymbol{m}} &= \boldsymbol{m} - \tilde{\boldsymbol{m}} \\
&= \boldsymbol{m}_\infty + \boldsymbol{m}_s - (\tilde{\boldsymbol{m}})_\infty - (\tilde{\boldsymbol{m}})_s \\
&= \boldsymbol{m}_s - (\tilde{\boldsymbol{m}})_s \\
&= \boldsymbol{m}_s - (\boldsymbol{m}_s)^\sim + (\boldsymbol{m}_s)^\sim - (\tilde{\boldsymbol{m}})_s.
\end{aligned}$$

If we are able to show that

$$(\boldsymbol{m}_s)^\sim = (\tilde{\boldsymbol{m}})_s \tag{8.15}$$

then the proof is complete because of the maximality of $(\boldsymbol{m}_s)^\sim$ in $\mathcal{N}(\boldsymbol{m}_s)$. In order to prove (8.15), note that $(\boldsymbol{m}_s)_\infty$ is identically zero on $\mathcal{T}^\diamond$ because of

$$\boldsymbol{m} = \boldsymbol{m}_\infty + (\boldsymbol{m}_s)_\infty + (\boldsymbol{m}_s)_s$$

and the maximality of $\boldsymbol{m}_\infty$ in $\mathcal{N}(\boldsymbol{m})$. Similarly, $((\boldsymbol{m}_s)^\sim)_\infty$ is the zero measure on $\mathcal{T}^\diamond$. Observe that

$$\boldsymbol{m} - (\boldsymbol{m}_s)^\sim = \boldsymbol{m}_\infty + (\boldsymbol{m}_s - (\boldsymbol{m}_s)^\sim),$$

the T_s-measure on the right hand side being monotone on $\mathcal{T}$. Since $\tilde{\boldsymbol{m}}$ is the greatest element in $\mathcal{N}(\boldsymbol{m})$, we have $\tilde{\boldsymbol{m}} \gg (\boldsymbol{m}_s)^\sim$. From the monotonicity of $\mathring{\boldsymbol{m}} = \boldsymbol{m}_s - (\tilde{\boldsymbol{m}})_s$ and the maximality of $(\boldsymbol{m}_s)^\sim$ in $\mathcal{N}(\boldsymbol{m}_s)$, we deduce that $(\boldsymbol{m}_s)^\sim - (\tilde{\boldsymbol{m}})_s$ is monotone and, subsequently, $(\boldsymbol{m}_s)^\sim \geq (\tilde{\boldsymbol{m}})_s$. Assume that $(\boldsymbol{m}_s)^\sim \neq (\tilde{\boldsymbol{m}})_s$, and define the monotone T_∞-measure $\overline{\boldsymbol{m}} : \mathcal{T}^\diamond \longrightarrow [0, \infty[$ by

$$\overline{\boldsymbol{m}} = (\tilde{\boldsymbol{m}})_\infty + (\boldsymbol{m}_s)^\sim.$$

Then we have

$$\begin{aligned}
\overline{\boldsymbol{m}} - \tilde{\boldsymbol{m}} &= (\tilde{\boldsymbol{m}})_\infty + (\boldsymbol{m}_s)^\sim - \tilde{\boldsymbol{m}} \\
&= (\tilde{\boldsymbol{m}})_\infty + (\tilde{\boldsymbol{m}})_s - (\tilde{\boldsymbol{m}})_s + (\boldsymbol{m}_s)^\sim - \tilde{\boldsymbol{m}} \qquad (8.16) \\
&= (\boldsymbol{m}_s)^\sim - (\tilde{\boldsymbol{m}})_s
\end{aligned}$$

the latter being monotone, showing that $\overline{\boldsymbol{m}} \gg \tilde{\boldsymbol{m}}$. On the other hand, Claim 3 yields

$$\boldsymbol{m} - \overline{\boldsymbol{m}} = \boldsymbol{m}_s - (\boldsymbol{m}_s)^{\sim},$$

which implies that $\boldsymbol{m} - \overline{\boldsymbol{m}}$ is monotone too, i.e., $\overline{\boldsymbol{m}} \in \mathcal{N}(\boldsymbol{m})$. Therefore, $\overline{\boldsymbol{m}} = \tilde{\boldsymbol{m}}$ which, together with (8.16), contradicts the assumption $(\boldsymbol{m}_s)^{\sim} \neq (\tilde{\boldsymbol{m}})_s$, thus completing the proof of Claim 4.

Now (8.14) shows that $\boldsymbol{m} = \mathring{\boldsymbol{m}} + \tilde{\boldsymbol{m}}$, where $\mathring{\boldsymbol{m}}$ is monotonically irreducible and $\tilde{\boldsymbol{m}}$ is generated by definition. It remains to show that (8.8) holds. From Claim 1 we know that $(\tilde{\boldsymbol{m}})^{\vee} \leq \overset{\vee}{\boldsymbol{m}}$ and, therefore, $(\tilde{\boldsymbol{m}})^{\vee}$ is absolutely continuous with respect to $\overset{\vee}{\boldsymbol{m}}$. Thus we may assume that the Radon–Nikodym derivative

$$\frac{\mathrm{d}(\tilde{\boldsymbol{m}})^{\vee}}{\mathrm{d}\overset{\vee}{\boldsymbol{m}}}$$

has values in $[0,1]$ only. Taking into account (8.12) we obtain for each $A \in \mathcal{T}^{\diamond}$

$$(\tilde{\boldsymbol{m}})_s(A) = \int_{\{A>0\}} l \, \mathrm{d}(\tilde{\boldsymbol{m}})^{\vee} = \int_{\{A>0\}} l \cdot \frac{\mathrm{d}(\tilde{\boldsymbol{m}})^{\vee}}{\mathrm{d}\overset{\vee}{\boldsymbol{m}}} \, \mathrm{d}\overset{\vee}{\boldsymbol{m}} \, .$$

Writing

$$g = l \cdot \frac{\mathrm{d}(\tilde{\boldsymbol{m}})^{\vee}}{\mathrm{d}\overset{\vee}{\boldsymbol{m}}}$$

and taking into account Claim 3 and (8.1), we have for all $A \in \mathcal{T}^{\diamond}$

$$\begin{aligned}
\tilde{\boldsymbol{m}}\,(A) &= (\tilde{\boldsymbol{m}})_{\infty}(A) + (\tilde{\boldsymbol{m}})_s(A) \\
&= \boldsymbol{m}_{\infty}(A) + (\tilde{\boldsymbol{m}})_s(A) \\
&= \int_{\{A>0\}} (1-f) \cdot A \, \mathrm{d}\overset{\vee}{\boldsymbol{m}} + \int_{\{A>0\}} g \, \mathrm{d}\overset{\vee}{\boldsymbol{m}} \\
&= \int_{\{A>0\}} [g + (1-f) \cdot A] \, \mathrm{d}\overset{\vee}{\boldsymbol{m}} \, .
\end{aligned}$$

Putting $h = 1 - f$ gives (8.8). Since $\tilde{\boldsymbol{m}}$ is unique and l, f and the Radon–Nikodym derivative

$$\frac{\mathrm{d}(\tilde{\boldsymbol{m}})^{\vee}}{\mathrm{d}\overset{\vee}{\boldsymbol{m}}}$$

are $\overset{\vee}{\boldsymbol{m}}$-a.e. uniquely determined, so are g and h. □

8.7 Combining the results of the First Representation Theorem 5.8 and the Decomposition Theorem 8.6 leads to the following:

Corollary. *If $s \in]0, \infty]$, $\mathcal{T}$ is a T_s-tribe and $\boldsymbol{m}$ is a finite monotone T_s-measure on $\mathcal{T}$, then there exists a unique finite nonnegative σ-additive measure $\mathbf{p}$ on $\mathcal{T}^{\vee}$, a $\mathbf{p}$-a.e. uniquely determined $\mathcal{T}^{\vee}$-Markov–kernel $\mathfrak{K} : X \times \mathcal{B}_1 \longrightarrow \mathbb{R}$ and a unique monotonically irreducible T_s-measure $\mathring{\boldsymbol{m}}$ on $\mathcal{T}$ such that for all $A \in \mathcal{T}$*

$$\boldsymbol{m}(A) = \mathring{\boldsymbol{m}}\,(A) + \int_X \mathfrak{K}(x, [0, A(x)[)\,\mathrm{d}\mathbf{p}(x).$$

8.8 Remark. From the results in [MESIAR, to appear] it follows that for $s \in]0, \infty]$ each T_s-measure $\boldsymbol{m}$, which is defined on T_s-tribe on a countable set, is a generated measure, i.e., $\mathring{\boldsymbol{m}}$ equals the zero measure (cf. Remark 3.4(iii)).

Chapter III

T_∞–Measures

9 Jordan Decomposition of Bounded T_∞–Measures

9.1 It was shown in Chapter II (see Theorem 8.6) that *monotone* fundamental triangular norm based measures can be decomposed into a sum consisting of a generated measure and a monotonically irreducible measure. Moreover, if the tribe on which the fundamental triangular norm based measure is defined is a *generated* one, then the monotonically irreducible component is necessarily zero (cf. Theorem 7.1).

It is an interesting open question whether the monotonically irreducible component of a fundamental triangular norm based monotone measure on a *nongenerated* tribe is always identically zero. We shall prove that in the particular case of the T_∞-measures, this happens even if the T_∞-measure is *not monotone.* This fact is of essential interest in many game theoretical considerations in Chapters IV and V, and it is a direct consequence of the "Jordan decomposability" of the T_∞-measures proven in this and in subsequent sections.

The fact that T_∞-measures are Jordan decomposable can be deduced directly from a result in [SCHMIDT 1982]. However, we present the proof by BUTNARIU [1986] because some technical details of this proof will be used later.

The Jordan decomposability of T_∞-measures defined on T_∞-tribes is used in Sections 12 — 14 in order to prove an extension of a theorem by DVORETZKY, WALD & WOLFOWITZ [1951] which, in turn, is a generalization of the classical Liapounoff theorem ([LIAPOUNOFF 1940]) concerning the compactness and convexity of the range of vector measures. This result is one of the basic tools required in Chapter IV to show the existence of an *Aumann–Shapley* value on the space of games with fuzzy coalitions spanned by positive integer powers of monotone T_∞-measures and, implicitly, for the extrapolation of the Aumann–Shapley value from the space we shall call *pFNA* to spaces of games which are much larger than *pFNA* itself in Chapter V.

It is natural to ask in which sense T_∞-measures differ from other fundamental triangular norm-based measures. In order to answer this question, we recall that T-measures are always T-countably additive functions but, in general, T-countable additivity does not characterize T-measures (see Remark 5.5). However, T_∞-countably additive functions are necessarily T_∞-measures. To be precise, we have the following result:

Proposition. *If* $\boldsymbol{m}$ *is a finite* T_∞*-countably additive function on a* T_∞*-tribe* $\mathcal{T}$*, then* $\boldsymbol{m}$ *is a* T_∞*-measure on* $\mathcal{T}$*.*

PROOF: Observe that

$$(A\,\mathbf{T}_\infty\,(A\,\mathbf{T}_\infty\,B)^c)\,\mathbf{T}_\infty\,B = \emptyset,$$

$$A\,\mathbf{S}_\infty\,B = (A\,\mathbf{T}_\infty\,(A\,\mathbf{T}_\infty\,B)^c)\,\mathbf{S}_\infty\,B.$$

Therefore, using the T_∞-additivity of $\boldsymbol{m}$ we get

$$\boldsymbol{m}(A\,\mathbf{S}_\infty\,B) = \boldsymbol{m}(A\,\mathbf{T}_\infty\,(A\,\mathbf{T}_\infty\,B)^c) + \boldsymbol{m}(B).$$

But

$$A = (A\,\mathbf{T}_\infty\,(A\,\mathbf{T}_\infty\,B)^c)\,\mathbf{S}_\infty\,(A\,\mathbf{T}_\infty\,B),$$

and the fuzzy sets $A\,\mathbf{T}_\infty\,(A\,\mathbf{T}_\infty\,B)^c$ and $A\,\mathbf{T}_\infty\,B$ are T_∞-disjoint. Hence,

$$\boldsymbol{m}(A) = \boldsymbol{m}(A\,\mathbf{T}_\infty\,(A\,\mathbf{T}_\infty\,B)^c) + \boldsymbol{m}(A\,\mathbf{T}_\infty\,B),$$

and this implies

$$\boldsymbol{m}(A\,\mathbf{S}_\infty\,B) = \boldsymbol{m}(A) - \boldsymbol{m}(A\,\mathbf{T}_\infty\,B) + \boldsymbol{m}(B),$$

showing that $\boldsymbol{m}$ is a T_∞-valuation on $\mathcal{T}$. Finally, we show that T_∞-countably additive functions are left continuous. Let $(A_n)_{n\in\mathbb{N}}$ be a nondecreasing sequence in $\mathcal{T}$ which converges to A. Define $A_0 = \emptyset$ and for each $n \in \mathbb{N}$

$$B_n = A_n\,\mathbf{T}_\infty\,A_{n-1}^c.$$

Then for all $n \in \mathbb{N}$

$$(\mathop{\mathbf{S}_\infty}_{i=1}^{n} B_i)\,\mathbf{T}_\infty\,B_{n+1} = A_n\,\mathbf{T}_\infty\,(A_{n+1}\,\mathbf{T}_\infty\,A_n^c) = \emptyset,$$

showing that the sequence $(B_n)_{n\in\mathbb{N}}$ is T_∞-disjoint. Since for each $n \in \mathbb{N}$ we have

$$A_n = \mathop{\mathbf{S}_\infty}_{k=1}^{n} B_k,$$

we obtain

$$A = \mathop{\mathbf{S}_\infty}_{n=1}^{\infty} B_n,$$

and, consequently,

$$\begin{aligned}
\lim_{n\to\infty} \boldsymbol{m}(A_n) &= \lim_{n\to\infty} \boldsymbol{m}(\mathbf{S}_{\infty}{}_{k=1}^{n} B_k) \\
&= \lim_{n\to\infty} \sum_{k=1}^{n} \boldsymbol{m}(B_k) \\
&= \sum_{n=1}^{\infty} \boldsymbol{m}(B_n) \\
&= \boldsymbol{m}(\mathbf{S}_{\infty}{}_{n=1}^{\infty} B_n) \\
&= \boldsymbol{m}(A).
\end{aligned}$$

Hence, $\boldsymbol{m}$ is a T_∞-measure. □

9.2 Proposition 9.1 allows us to approach T_∞-measures on T_∞-tribes in a way similar to the treatment of (σ-additive) measures on σ-algebras of subsets of X.

Definition. Let $\mathcal{T}$ be a T_∞-tribe, $\boldsymbol{m} : \mathcal{T} \longrightarrow [-\infty, +\infty]$ a function such that $\boldsymbol{m}(\emptyset) = 0$, and $A \in \mathcal{T}$.

(i) The function $\boldsymbol{m}^+ : \mathcal{T} \longrightarrow [0, \infty]$ defined by

$$\boldsymbol{m}^+(A) = \sup\{\boldsymbol{m}(B) \mid B \in \mathcal{T}(A)\} \tag{9.1}$$

is called *upper variation* of $\boldsymbol{m}$, where $\mathcal{T}(A)$ denotes the class of all fuzzy subsets of A contained in $\mathcal{T}$, i.e.,

$$\mathcal{T}(A) = \{B \mid B \in \mathcal{T}, B \leq A\}.$$

(ii) The function $\boldsymbol{m}^- : \mathcal{T} \longrightarrow [0, \infty]$ defined by

$$\boldsymbol{m}^-(A) = \sup\{-\boldsymbol{m}(B) \mid B \in \mathcal{T}(A)\} \tag{9.2}$$

is called *lower variation* of $\boldsymbol{m}$.

(iii) The function $\boldsymbol{m}^* : \mathcal{T} \longrightarrow [0, \infty]$ defined by

$$\boldsymbol{m}^* = \boldsymbol{m}^+ + \boldsymbol{m}^-$$

is called *total variation* of $\boldsymbol{m}$.

Obviously, these functions are all monotone, and they satisfy

$$\begin{aligned} \boldsymbol{m}^+ &= (-\boldsymbol{m})^-, \\ \boldsymbol{m}^- &= (-\boldsymbol{m})^+, \end{aligned} \tag{9.3}$$

$$\boldsymbol{m}^+(\emptyset) = \boldsymbol{m}^-(\emptyset) = \boldsymbol{m}^*(\emptyset) = 0.$$

To simplify notation, from now on we shall use the following shortcut: for fuzzy sets $A, B \in \mathscr{T}$ we shall write

$$A \ominus B = A\,\mathbf{T}_\infty\, B^c.$$

Obviously, each T_∞-tribe $\mathscr{T}$ is closed under this operation.

9.3 Proposition.

(i) If $\boldsymbol{m}$ is finitely T_∞-additive, then $\boldsymbol{m}^+, \boldsymbol{m}^-$ and $\boldsymbol{m}^*$ are finitely T_∞-additive too.

(ii) If $\boldsymbol{m}$ is a T_∞-measure, then $\boldsymbol{m}^+, \boldsymbol{m}^-$ and $\boldsymbol{m}^*$ are monotone T_∞-measures.

PROOF: When proving (i), it is sufficient to show that $\boldsymbol{m}^+$ is T_∞-additive (cf. (9.3)). Let E and F be T_∞-disjoint fuzzy sets in $\mathscr{T}$. For each $B \in \mathscr{T}(E\,\mathbf{S}_\infty\, F)$ the fuzzy sets $C = B\,\mathbf{T}_\infty\, E$ and $D = B \ominus C$ are elements of $\mathscr{T}$, and they form a T_∞-partition of B. Since $D \in \mathscr{T}(E)$ and $C \in \mathscr{T}(F)$, we have

$$\boldsymbol{m}(B) = \boldsymbol{m}(C) + \boldsymbol{m}(D) \leq \boldsymbol{m}^+(F) + \boldsymbol{m}^+(E)$$

which implies

$$\boldsymbol{m}^+(E\,\mathbf{S}_\infty\, F) \leq \boldsymbol{m}^+(E) + \boldsymbol{m}^+(F).$$

For each $\varepsilon > 0$ there exist fuzzy sets $A \in \mathscr{T}(E)$ and $B \in \mathscr{T}(F)$ such that

$$\begin{aligned} \boldsymbol{m}^+(E) &\leq \boldsymbol{m}(A) + \frac{\varepsilon}{2}, \\ \boldsymbol{m}^+(F) &\leq \boldsymbol{m}(B) + \frac{\varepsilon}{2}. \end{aligned}$$

From $A\,\mathbf{T}_\infty\, B \leq E\,\mathbf{T}_\infty\, F$ we deduce

$$\boldsymbol{m}^+(E) + \boldsymbol{m}^+(F) \leq \boldsymbol{m}(A) + \boldsymbol{m}(B) + \varepsilon = \boldsymbol{m}(A\,\mathbf{S}_\infty\, B) + \varepsilon \leq \boldsymbol{m}^+(E\,\mathbf{S}_\infty\, F) + \varepsilon.$$

Taking the limit $\varepsilon \to 0$ we obtain

$$\boldsymbol{m}^+(E) + \boldsymbol{m}^+(F) \leq \boldsymbol{m}^+(E\,\mathbf{S}_\infty\, F),$$

and this completes the proof of (i).

In order to show the validity of (ii), assume that $\boldsymbol{m}$ is a T_∞-measure. According to Proposition 9.1 and equations (9.3), it is sufficient to prove that $\boldsymbol{m}^+$ is T_∞-countably additive. Let $(A_k)_{k\in\mathbb{N}}$ be a T_∞-disjoint sequence of fuzzy sets in $\mathscr{T}$. Then the sequence $(B_n)_{n\in\mathbb{N}}$ defined by

$$B_n = \mathop{\mathbf{S}_\infty}_{k=1}^{n} A_k$$

is nondecreasing and, since $\boldsymbol{m}^+$ is T_∞-additive, it follows that for each $n \in \mathbb{N}$

$$\boldsymbol{m}^+(B_n) = \sum_{k=1}^{n} \boldsymbol{m}^+(A_k).$$

Thus we have

$$\lim_{n\to\infty} \boldsymbol{m}^+(B_n) = \sum_{k=1}^{\infty} \boldsymbol{m}^+(A_k).$$

Since for each $n \in \mathbb{N}$

$$B_n \le \mathop{\mathbf{S}_\infty}_{k=1}^{\infty} A_k$$

and since $\boldsymbol{m}^+$ is monotone, we obtain

$$\boldsymbol{m}^+(\mathop{\mathbf{S}_\infty}_{k=1}^{\infty} A_k) \ge \sum_{k=1}^{\infty} \boldsymbol{m}^+(A_k). \tag{9.4}$$

For each

$$A \in \mathscr{T}(\mathop{\mathbf{S}_\infty}_{k=1}^{\infty} A_k)$$

we have

$$A = A\,\mathbf{T}_0\,(\mathop{\mathbf{S}_\infty}_{k=1}^{\infty} A_k) = A\,\mathbf{T}_0\,(\mathop{\mathbf{S}_0}_{n=1}^{\infty} B_n) = \mathop{\mathbf{S}_0}_{n=1}^{\infty}(A\,\mathbf{T}_0\,B_n) = \lim_{n\to\infty}(A\,\mathbf{T}_0\,B_n).$$

The sequence $(A\,\mathbf{T}_0\,B_n)_{n\in\mathbb{N}}$ is nondecreasing and $\boldsymbol{m}$ is left–continuous since it is a T_∞-measure. Hence we have

$$\boldsymbol{m}(A) = \boldsymbol{m}(\lim_{n\to\infty}(A\,\mathbf{T}_0\,B_n)) = \lim_{n\to\infty} \boldsymbol{m}(A\,\mathbf{T}_0\,B_n) \le \lim_{n\to\infty} \boldsymbol{m}^+(B_n)$$

because of $A\,\mathbf{T}_0\,B_n \in \mathscr{T}(B_n)$, which implies

$$\boldsymbol{m}(A) \le \sum_{k=1}^{\infty} \boldsymbol{m}^+(A_k).$$

Therefore, we obtain equality even in (9.4), and the proposition is proved. □

9.4 Proposition.

(i) A finitely T_∞-additive function $\boldsymbol{m} : \mathscr{T} \longrightarrow [-\infty, +\infty]$ is bounded if and only if $\boldsymbol{m}^*$ is bounded.

(ii) If the finitely T_∞-additive function $\boldsymbol{m} : \mathscr{T} \longrightarrow [-\infty, +\infty]$ is bounded, then we have

$$\boldsymbol{m} = \boldsymbol{m}^+ - \boldsymbol{m}^-, \tag{9.5}$$

i.e., the pair $(\boldsymbol{m}^+, \boldsymbol{m}^-)$ is a Jordan decomposition of $\boldsymbol{m}$.

PROOF: We first show (i). If $\boldsymbol{m}^*$ is bounded, then for each $A \in \mathscr{T}$

$$|\boldsymbol{m}(A)| \leq \sup\{\boldsymbol{m}^*(B) \mid B \in \mathscr{T}\} < \infty,$$

i.e., $\boldsymbol{m}$ is bounded. Conversely, if $\boldsymbol{m}$ is bounded, then

$$\max(\boldsymbol{m}^+(A), \boldsymbol{m}^-(A)) \leq \sup\{|\boldsymbol{m}(B)| \mid B \in \mathscr{T}(A)\} \leq \sup\{|\boldsymbol{m}(B)| \mid B \in \mathscr{T}\} < \infty$$

for all $A \in \mathscr{T}$. Hence, $\boldsymbol{m}^+$ and $\boldsymbol{m}^-$ are both bounded, and, consequently, so is $\boldsymbol{m}^*$.

In order to prove (ii), assume that $\boldsymbol{m}$ is bounded. Then $\boldsymbol{m}^+, \boldsymbol{m}^-$ and $\boldsymbol{m}^*$ are bounded too, and for $A \in \mathscr{T}$ and $M \in \mathscr{T}(A)$ we have

$$\boldsymbol{m}(M) = \boldsymbol{m}(A) - \boldsymbol{m}(A \ominus M).$$

Since $A \ominus M \in \mathscr{T}(A)$, it follows that $\boldsymbol{m}^+(A) \geq \boldsymbol{m}(A \ominus M)$ and $\boldsymbol{m}^-(A) \geq -\boldsymbol{m}(A \ominus M)$. Hence,

$$\boldsymbol{m}(A) - \boldsymbol{m}^+(A) \leq \boldsymbol{m}(A) - \boldsymbol{m}(A \ominus M) \leq \boldsymbol{m}(A) + \boldsymbol{m}^-(A)$$

and, consequently,

$$\boldsymbol{m}(A) - \boldsymbol{m}^+(A) \leq \boldsymbol{m}(M) \leq \boldsymbol{m}(A) + \boldsymbol{m}^-(A).$$

From (9.1) and (9.2) we deduce

$$\boldsymbol{m}^-(A) \leq \boldsymbol{m}^+(A) - \boldsymbol{m}(A),$$
$$\boldsymbol{m}^+(A) \leq \boldsymbol{m}(A) + \boldsymbol{m}^-(A),$$

and this completes the proof. □

9.5 A combination of Proposition 9.4, Theorem 6.2 and Theorem 3.2 yields the following.

Corollary. *Let $\boldsymbol{m}$ be a bounded T_∞-measure on a T_∞-tribe $\mathscr{T}$. Then we have:*

(i) $\boldsymbol{m}$ is continuous in the sense of (6.1);

(ii) for all $A \in \mathscr{T}$,

$$\boldsymbol{m}(A) = \int_X A(x)\,\mathrm{d}((\boldsymbol{m}^+)^\vee - (\boldsymbol{m}^-)^\vee), \tag{9.6}$$

where $(\boldsymbol{m}^+)^\vee$ and $(\boldsymbol{m}^-)^\vee$ are the restrictions of $\boldsymbol{m}^+$ and $\boldsymbol{m}^-$ to $\mathscr{T}^\vee$, respectively;

(iii) $(\boldsymbol{m}^+)^\vee$ and $(\boldsymbol{m}^-)^\vee$ are the upper and lower variation of $\check{\boldsymbol{m}}$, the restriction of $\boldsymbol{m}$ to $\mathscr{T}^\vee$, and, therefore,

$$\check{\boldsymbol{m}} = (\boldsymbol{m}^+)^\vee - (\boldsymbol{m}^-)^\vee. \tag{9.7}$$

PROOF: Indeed, (i) follows from Proposition 6.1(ii), (ii) follows from Theorem 6.2, and (iii) is a consequence of (ii), taking into account Proposition 9.4. □

10 Jordan Decomposition of Finite T_∞-Measures

10.1 In Section 9 we have shown that bounded T_∞-measures on T_∞-tribes have Jordan decompositions and, therefore, are continuous (in the sense of (6.1)) and can be represented by integrals (see (9.6)). The aim of this section is to show that *finite* T_∞-measures on T_∞-tribes are necessarily bounded and, consequently, have Jordan decompositions by finite T_∞-measures.

Lemma. *Let $\boldsymbol{m}$ be a finite T_∞-measure on a T_∞-tribe $\mathscr{T}$. The following are equivalent:*

(i) $\boldsymbol{m}$ is bounded;

(ii) $\boldsymbol{m}^+(X) < \infty$ and $\boldsymbol{m}^-(X) < \infty$.

PROOF: The implication (i) ⇒ (ii) is trivial. Assume that (ii) holds. The monotonicity of $\boldsymbol{m}^+$ and $\boldsymbol{m}^-$ implies that they are both bounded. Thus $\boldsymbol{m}^*$ and, consequently, $\boldsymbol{m}$ are bounded (cf. Proposition 9.4). □

10.2 Theorem. *([BUTNARIU 1986]) If $\boldsymbol{m}$ is a finite T_∞-measure on a T_∞-tribe $\mathscr{T}$, then there exist nonnegative finite T_∞-measures $\boldsymbol{m}^+$ and $\boldsymbol{m}^-$ on $\mathscr{T}$ such that*

$$\boldsymbol{m} = \boldsymbol{m}^+ - \boldsymbol{m}^-, \tag{10.1}$$

i.e., $(\boldsymbol{m}^+, \boldsymbol{m}^-)$ is a Jordan decomposition of $\boldsymbol{m}$.

PROOF: Let $\boldsymbol{m}$ be a finite T_∞-measure on $\mathscr{T}$. According to Proposition 9.4 and Lemma 10.1, it suffices to show that $\boldsymbol{m}^+(X) < \infty$ and $\boldsymbol{m}^-(X) < \infty$. In order to show $\boldsymbol{m}^+(X) < \infty$, a fuzzy set $M \in \mathscr{T}$ such that $\boldsymbol{m}^+(M) = \infty$ will be called *unbounded*.

Claim: If M is unbounded, then there exists a nonincreasing sequence $(A_n)_{n\in\mathbb{N}}$ of unbounded fuzzy sets in $\mathscr{T}(M)$ such that $\boldsymbol{m}(A_n) \geq n$ for all $n \in \mathbb{N}$.

Let M be unbounded and assume that there exists an $n_0 \in \mathbb{N}$ such that $\boldsymbol{m}(A) < n_0$ for all unbounded fuzzy sets $A \in \mathscr{T}(M)$. Since M is unbounded, there exists an $M_1 \in \mathscr{T}(M)$ such that $\boldsymbol{m}(M_1) \leq n_0$, which means that M_1 is not unbounded, i.e.,

$$\sup\{\boldsymbol{m}(A) \mid A \in \mathscr{T}(M_1)\} < \infty. \tag{10.2}$$

Choose an arbitrary $B \in \mathscr{T}(M)$ and denote $B_1 = B\,\mathbf{T}_0\,M_1$, $B_2 = B \ominus B_1$. It is clear that $B_1\,\mathbf{T}_\infty\,B_2 = \emptyset$, $B_1\,\mathbf{S}_\infty\,B_2 = B$, $B_1 \in \mathscr{T}(M_1)$ and $B_2 \in \mathscr{T}(M \ominus M_1)$. Therefore,

$$\boldsymbol{m}(B) = \boldsymbol{m}(B_1) + \boldsymbol{m}(B_2)$$

and

$$\boldsymbol{m}^+(M \ominus M_1) \geq \boldsymbol{m}(B_2) = \boldsymbol{m}(B) - \boldsymbol{m}(B_1) \geq \boldsymbol{m}(B) - \boldsymbol{m}^+(M_1).$$

Since this is true for all $B \in \mathscr{T}(M)$, since M is unbounded and since (10.2) holds, we have

$$\boldsymbol{m}^+(M \ominus M_1) \geq \boldsymbol{m}^+(M) - \boldsymbol{m}^-(M_1) = \infty,$$

i.e., $M \ominus M_1$ is unbounded. Then, by assumption, $\boldsymbol{m}(M \ominus M_1) < n_0$, and there exists a fuzzy set $C_1 \in \mathscr{T}(M \ominus M_1)$ such that $\boldsymbol{m}(C_1) \geq 1$. Denote $M_2 = M_1\,\mathbf{S}_\infty\,C_1$. Then

$$M_1\,\mathbf{T}_\infty\,C_1 \leq M_1\,\mathbf{T}_\infty\,(M \ominus M_1) = \emptyset$$

and

$$\boldsymbol{m}(M_2) = \boldsymbol{m}(M_1) + \boldsymbol{m}(C_1) \geq n_0 + 1. \tag{10.3}$$

Using the assumption again yields that M_2 is not unbounded. Similar reasoning as above shows that $M \ominus M_2$ is unbounded too. Hence, $\boldsymbol{m}(M \ominus M_2) < n_0$ and $\boldsymbol{m}(C_2) \geq 1$ for some $C_2 \in \mathscr{T}(M \ominus M_2)$. Denote $M_3 = M_2\,S_\infty\,C_2$. Since

$$M_2\,\mathbf{T}_\infty\,C_2 \leq M_2\,\mathbf{T}_\infty\,(M \ominus M_2) = \emptyset$$

it follows from (10.3) that

$$\boldsymbol{m}(M_3) = \boldsymbol{m}(M_2) + \boldsymbol{m}(C_2) \geq n_0 + 2,$$

i.e., M_3 is not unbounded. In this way we obtain fuzzy sets $M_1, M_2, M_3, \ldots$ in $\mathscr{T}(M)$, none of which are unbounded, and $C_1, C_2, C_3, \ldots$ such that $C_n \in \mathscr{T}(M \ominus M_n)$ and $\boldsymbol{m}(C_n) \geq 1$ for all $n \in \mathbb{N}$. Moreover, we can show that for all $n \in \mathbb{N}$

$$\sum_{i=1}^{n} C_i \leq M - M_1. \tag{10.4}$$

Indeed, if $n = 1$ this is true because of the definition of C_1. If (10.4) holds for all $k \in \{1, 2, \ldots, n\}$, then we have for each $x \in X$

$$\begin{aligned} C_{n+1}(x) &\le \max(0, M(x) - M_{n+1}(x)) \\ &= M(x) - M_{n+1}(x) \\ &= M(x) - (M_n(x) + C_n(x)) \\ &= M(x) - (M_{n-1}(x) + C_{n-1}(x) + C_n(x)) \\ &\vdots \\ &= M(x) - M_1(x) - \sum_{i=1}^{n} C_i(x). \end{aligned}$$

Hence, (10.4) is proved by induction. An immediate consequence of (10.4) is that the sequence $(C_n)_{n\in\mathbb{N}}$ is T_∞-disjoint which, together with $\boldsymbol{m}(C_n) \ge 1$, yields

$$\boldsymbol{m}(\mathbf{S}_{\infty\, n=1}^{\infty} C_n) = \sum_{n=1}^{\infty} \boldsymbol{m}(C_n) = \infty.$$

Since $\boldsymbol{m}$ is finite, this contradicts the assumption that $\boldsymbol{m}(A) < n_0$ for all unbounded fuzzy sets $A \in \mathcal{T}(M)$. Therefore, there exists an unbounded fuzzy set $A_1 \in \mathcal{T}(M)$ such that $\boldsymbol{m}(A_1) \ge 1$. Since A_1 is unbounded, there exists an unbounded fuzzy set $A_2 \in \mathcal{T}(A_1)$ such that $\boldsymbol{m}(A_2) \ge 2$, and so on, showing that the claim holds.

To complete the proof of the theorem, assume that X is unbounded and choose the nonincreasing sequence $(A_n)_{n\in\mathbb{N}}$ of unbounded fuzzy sets in $\mathcal{T}$ which exists according to the claim. For each $n \in \mathbb{N}$ denote

$$B_n = \mathbf{S}_{\infty\, k=1}^{n-1} (A_k \ominus A_{k+1}).$$

Clearly, since $(A_n)_{n\in\mathbb{N}}$ is nonincreasing, the sequence $(A_k \ominus A_{k+1})_{k\in\mathbb{N}}$ is T_∞-disjoint, and therefore,

$$\boldsymbol{m}(B_n) = \sum_{k=1}^{n-1} \boldsymbol{m}(A_k \ominus A_{k+1}) = \boldsymbol{m}(A_1) - \boldsymbol{m}(A_n).$$

The sequence $(B_n)_{n\in\mathbb{N}}$ is nondecreasing which leads to

$$\boldsymbol{m}(\lim_{n\to\infty} B_n) = \lim_{n\to\infty} \boldsymbol{m}(B_n) = \boldsymbol{m}(A_1) - \lim_{n\to\infty} \boldsymbol{m}(A_n) = -\infty,$$

where the latter equality follows from $\boldsymbol{m}(A_n) \ge n$ and $\boldsymbol{m}(A_1) < \infty$. This contradicts the finiteness of $\boldsymbol{m}$. Therefore, the assumption that X is unbounded is false and, consequently, $\boldsymbol{m}^+(X) < \infty$.

The inequality $\boldsymbol{m}^-(X) < \infty$ is shown in complete analogy, $\boldsymbol{m}$ is bounded by Lemma 10.1, and the Jordan decomposition exists because of Proposition 9.4. □

10.3 Since each bounded T_∞-measure is automatically finite, in the proof of Theorem 10.2 we have implicitly proven the following result:

Corollary. *A T_∞-measure $\boldsymbol{m}$ on a T_∞-tribe $\mathscr{T}$ is bounded if and only if it is finite.*

10.4 Combining Corollary 9.4 and Theorem 10.2 we realize that finite T_∞-measures on T_∞-tribes are exactly (Lebesgue) integrals:

Corollary. *Let $\boldsymbol{m}$ be a finite T_∞-measure on a T_∞-tribe $\mathscr{T}$. Then we have*

(i) $\boldsymbol{m}$ is continuous in the sense of (6.1);

(ii) for all $A \in \mathscr{T}$

$$\boldsymbol{m}(A) = \int_X A(x)\,\mathrm{d}(\check{\boldsymbol{m}}^+ - \check{\boldsymbol{m}}^-), \tag{10.5}$$

where $\check{\boldsymbol{m}}^+$ and $\check{\boldsymbol{m}}^-$ are the restrictions of $\boldsymbol{m}^+$ and $\boldsymbol{m}^-$ to $\mathscr{T}^\vee$, respectively;

(iii) $\check{\boldsymbol{m}}^+$ and $\check{\boldsymbol{m}}^-$ are the upper and the lower variation, respectively, of , the restriction of $\check{\boldsymbol{m}}$ to $\mathscr{T}^\vee$, and

$$\check{\boldsymbol{m}} = \check{\boldsymbol{m}}^+ - \check{\boldsymbol{m}}^-; \tag{10.6}$$

(iv) if (P, Q) is a Hahn decomposition of X with respect to the σ-additive measure $\check{\boldsymbol{m}}$ on $\mathscr{T}^\vee$, then for all $A \in \mathscr{T}$

$$\boldsymbol{m}(Q) = -\boldsymbol{m}^-(Q) \leq \boldsymbol{m}(A) \leq \boldsymbol{m}^+(P) = \boldsymbol{m}(P). \tag{10.7}$$

11 Absolute Continuity of T_∞-Measures

11.1 T_∞-measures on not necessarily generated T_∞-tribes will play an essential role in the following exposition. Some of the important properties of such T_∞-measures are the absolute continuity and the singularity of T_∞-measures. These will be studied in this section.

Definition. Let $\boldsymbol{m}$ and $\boldsymbol{n}$ be T_∞-measures on a T_∞-tribe $\mathscr{T}$. The measure $\boldsymbol{n}$ is called $\boldsymbol{m}$*-absolutely continuous* (briefly $\boldsymbol{n} \ll \boldsymbol{m}$) if for each $A \in \mathscr{T}$

$$\boldsymbol{n}(A) = 0 \quad \text{whenever} \quad \boldsymbol{m}^*(A) = 0. \tag{11.1}$$

Absolute continuity is a reflexive and transitive relation. Moreover, if $\boldsymbol{n}_1 \ll \boldsymbol{m}$ and $\boldsymbol{n}_2 \ll \boldsymbol{m}$, then $\boldsymbol{n}_1 \pm \boldsymbol{n}_2 \ll \boldsymbol{m}$.

11.2 There are several characterizations of the absolute continuity:

Lemma. *Let $\boldsymbol{m}$ and $\boldsymbol{n}$ be T_∞-measures on a T_∞-tribe $\mathscr{T}$. The following are equivalent:*

(i)

$$\boldsymbol{n} \ll \boldsymbol{m}; \tag{11.2}$$

(ii) for all $A \in \mathscr{T}$

$$\boldsymbol{n}(A) = 0 \quad \textit{whenever} \quad \boldsymbol{m}^*(A) = 0; \tag{11.3}$$

(iii) for all $A \in \mathscr{T}$

$$\boldsymbol{n}^+(A) = 0 = \boldsymbol{n}^-(A) \quad \textit{whenever} \quad \boldsymbol{m}^*(A) = 0; \tag{11.4}$$

(iv) for all $A \in \mathscr{T}$

$$\boldsymbol{n}^*(A) = 0 \quad \textit{whenever} \quad \boldsymbol{m}^*(A) = 0; \tag{11.5}$$

(v)

$$\boldsymbol{n}^* \ll \boldsymbol{m}^*; \tag{11.6}$$

(vi) for each $\varepsilon > 0$, there exists a $\delta_\varepsilon > 0$ such that for each $A \in \mathscr{T}$

$$\boldsymbol{n}^*(A) < \varepsilon \quad \textit{whenever} \quad \boldsymbol{m}^*(A) < \delta_\varepsilon; \tag{11.7}$$

(vii) for each $\varepsilon > 0$, there exists an $\eta_\varepsilon > 0$ such that for each $A \in \mathscr{T}$

$$|\boldsymbol{n}(A)| < \varepsilon \quad \textit{whenever} \quad \boldsymbol{m}^*(A) < \eta_\varepsilon; \tag{11.8}$$

(viii)

$$\check{\boldsymbol{n}} \ll \check{\boldsymbol{m}}; \tag{11.9}$$

(ix)

$$(\check{\boldsymbol{n}})^* \ll (\check{\boldsymbol{m}})^*. \tag{11.10}$$

PROOF: The equivalences

$$(\text{i}) \Leftrightarrow (\text{ii}) \Leftrightarrow (\text{iii}) \Leftrightarrow (\text{iv}) \Leftrightarrow (\text{v}),$$

$$(\text{vi}) \Leftrightarrow (\text{vii}),$$

$$(\text{viii}) \Leftrightarrow (\text{ix})$$

are trivial. The equivalence of (v) and (viii) follows from Corollary 10.4. We only have to show that (iv) and (vi) are equivalent. Assume that (vi) holds and choose $A \in \mathscr{T}$ such that $\boldsymbol{m}^*(A) = 0$. Then $\boldsymbol{m}^*(A) < \delta_\varepsilon$ and, as a consequence, $\boldsymbol{n}^*(A) < \varepsilon$ for each $\varepsilon > 0$. Hence, $\boldsymbol{n}^*(A) = 0$ and (iv) follows. Conversely, assuming (iv), suppose that there exists a real number $\varepsilon_0 > 0$ such that for each $\delta > 0$ there is an $A(\delta) \in \mathscr{T}$ with $\boldsymbol{n}^*(A(\delta)) \geq \varepsilon_0$ such that $\boldsymbol{m}^*(A(\delta)) < \delta$. For each $n \in \mathbb{N}$ denote $A_n = A(\frac{1}{2^n})$. Clearly the limit

$$A = \lim_{n\to\infty} \mathbf{S}_{\infty\, k=n}^{\infty} A_k$$

exists and $A \in \mathscr{T}$. Then

$$0 \leq \boldsymbol{m}^*(A) = \lim_{n\to\infty} \boldsymbol{m}^*(\mathbf{S}_{\infty\, k=n}^{\infty} A_k) \leq \sum_{k=n}^{\infty} \boldsymbol{m}^*(A_k) \leq \lim_{n\to\infty} \sum_{k=n}^{\infty} \frac{1}{2^k} = 0,$$

i.e., $\boldsymbol{m}^*(A) = 0$. On the other hand, we have

$$\boldsymbol{n}^*(A) = \lim_{n\to\infty} \boldsymbol{n}^*(\mathbf{S}_{\infty\, k=n}^{\infty} A_k) \geq \lim_{n\to\infty} \boldsymbol{n}^*(A_n) \geq \varepsilon_0 > 0,$$

i.e., $\boldsymbol{n}^*(A) \neq 0$, and this contradicts (iv). □

11.3 A property which is dual to absolute continuity is that of singularity:

Definition. Let $\boldsymbol{m}$ and $\boldsymbol{n}$ be T_∞-measures on a T_∞-tribe $\mathscr{T}$. The T_∞-measure $\boldsymbol{n}$ is said to be $\boldsymbol{m}$*-singular* (briefly $\boldsymbol{n} \perp \boldsymbol{m}$) if there exists a fuzzy set $A \in \mathscr{T}$ such that

$$\boldsymbol{n}^*(A) = \boldsymbol{m}^*(A^c) = 0.$$

11.4 There exist characterizations of the singularity which are dual to the characterizations of the absolute continuity and which are listed below.

Lemma. *Let $\boldsymbol{m}$ and $\boldsymbol{n}$ be measures on a T_∞-tribe $\mathscr{T}$. The following are equivalent:*

(i)

$$\boldsymbol{n} \perp \boldsymbol{m}; \tag{11.11}$$

(ii)

$$\boldsymbol{m} \perp \boldsymbol{n}; \tag{11.12}$$

(iii) there exists a fuzzy subset A of X such that $A \in \mathscr{T}$ and

$$\boldsymbol{m}^*(A) = \boldsymbol{n}^*(A^c) = 0; \tag{11.13}$$

(iv)

$$\boldsymbol{n}^+ \perp \boldsymbol{m} \quad \textit{and} \quad \boldsymbol{n}^- \perp \boldsymbol{m}; \tag{11.14}$$

(v)

$$\boldsymbol{n}^* \perp \boldsymbol{m}^*; \tag{11.15}$$

(vi)

$$\check{\boldsymbol{n}} \perp \check{\boldsymbol{m}}; \tag{11.16}$$

(vii)

$$(\check{\boldsymbol{n}})^* \perp (\check{\boldsymbol{m}})^*. \tag{11.17}$$

PROOF: Since all the other implications are trivial we only show that (i) implies (iii). If $\boldsymbol{m}^*(A^c) = \boldsymbol{n}^*(A) = 0$, then for $B = A^c$ we have $B^c = (A^c)^c$ and, consequently, $\boldsymbol{m}^*(B) = \boldsymbol{n}^*(B^c) = 0$. □

11.5 The following properties of the absolute continuity and singularity will be used later.

Proposition.

(i) If $\boldsymbol{n}_1, \boldsymbol{n}_2$ and $\boldsymbol{m}$ are finite T_∞-measures on a T_∞-tribe $\mathscr{T}$ such that $\boldsymbol{n}_1 \perp \boldsymbol{m}$ and $\boldsymbol{n}_2 \perp \boldsymbol{m}$, then $\boldsymbol{n}_1 \pm \boldsymbol{n}_2 \perp \boldsymbol{m}$.

(ii) If $\boldsymbol{m}$ and $\boldsymbol{n}$ are finite T_∞-measures on a T_∞-tribe $\mathscr{T}$ such that $\boldsymbol{n} \ll \boldsymbol{m}$ and $\boldsymbol{n} \perp \boldsymbol{m}$, then $\boldsymbol{n}$ equals the zero measure.

PROOF: The proof of (i) is trivial. In order to show (ii), let $\boldsymbol{m}$ and $\boldsymbol{n}$ be T_∞-measures on $\mathscr{T}$ such that $\boldsymbol{n} \ll \boldsymbol{m}$ and $\boldsymbol{n} \perp \boldsymbol{m}$. Since $\boldsymbol{n} \ll \boldsymbol{m}$, it follows that for each $k \in \mathbb{N}$ there is a $\delta_{1/k} > 0$ such that $\boldsymbol{n}^*(F) < 1/k$ whenever $F \in \mathscr{T}$ and $\boldsymbol{m}^*(F) < \delta_{1/k}$. Since $\boldsymbol{n} \perp \boldsymbol{m}$, there is a set $A \in \mathscr{T}$ such that $\boldsymbol{n}^*(A^c) = \boldsymbol{m}^*(A) = 0$. Clearly, $\boldsymbol{n}^*(A) = 0$ since $\boldsymbol{m}^*(A) < \delta_{1/k}$ for all $k \in \mathbb{N}$. If $E \in \mathscr{T}$,then

$$\boldsymbol{n}^*(E) = \boldsymbol{n}^*(E\,T_\infty\,A) + \boldsymbol{n}^*(E\,T_\infty\,A^c) = 0$$

because of $E\,T_\infty\,A \in \mathscr{T}(A), E\,T_\infty\,A^c \in \mathscr{T}(A^c)$ and the monotonicity of $\boldsymbol{n}^*$. □

12 Vector T_∞-Measures with Darboux Property

12.1 In Section 10 we proved that finite T_∞-measures on T_∞-tribes have bounded range. The question is now whether or under which conditions the range of a finite T_∞-measure on a T_∞-tribe (generated or not) is exactly an interval. This problem is studied here in a more general context, i.e., we ask whether or under which conditions a finite vector valued T_∞-measure on a T_∞-tribe has a convex and compact range. For measures defined on a σ-algebra of crisp subsets of X (which are particular T_∞-measures) this problem was approached by POPRUGENKO [1928], and it was solved by LIAPOUNOFF [1949]. HALMOS [1947, 1948], BLACKWELL [1951], CHERNOFF [1951] and LINDENSTRAUSS [1966] reproved and/or extended Liapounoff's result. In view of Corollary 10.4, DVORETZKI et al. [1951] have shown that the range of a vector T_∞-measure *on* a *generated* T_∞-tribe is necessarily convex and compact. We intend to generalize this result by showing that nonatomic T_∞-measures on T_∞-tribes (generated or not) have convex and compact range. To this end, we first study the class of T_∞-measures with Darboux property.

12.2 In this section $\mathscr{T}$ denotes a T_∞-tribe on X. In what follows, all T_∞-measures are assumed to be defined on $\mathscr{T}$. An *n-vector T_∞-measure on $\mathscr{T}$* is a function $\boldsymbol{m} = (\boldsymbol{m}_1, \boldsymbol{m}_2, \ldots, \boldsymbol{m}_n)$ from $\mathscr{T}$ to $\mathbb{R}^n$ such that each $\boldsymbol{m}_i$ is a finite T_∞-measure. If $\boldsymbol{m}$ is an n-vector T_∞-measure, we denote by

$$\mathcal{R}(\boldsymbol{m}) = \{\boldsymbol{m}(W) \mid W \in \mathscr{T}\}$$

the *range* of $\boldsymbol{m}$. Clearly, $0 \in \mathcal{R}(\boldsymbol{m})$ and $\mathcal{R}(\boldsymbol{m})$ is bounded in $\mathbb{R}^n$ (cf. Corollary 10.3).

Definition. An n-vector T_∞-measure $\boldsymbol{m} = (\boldsymbol{m}_1, \boldsymbol{m}_2, \ldots, \boldsymbol{m}_n)$ has the *Darboux property* if for each pair (E, F) of fuzzy sets in $\mathscr{T}$ with $E \le F$ there exists a family $(W_r)_{r\in[0,1]}$ in $\mathscr{T}$ such that for all $r \in [0,1]$

$$E \le W_r \le F; \tag{12.1}$$

$$\boldsymbol{m}(W_r) = (1-r)\cdot \boldsymbol{m}(E) + r \cdot \boldsymbol{m}(F). \tag{12.2}$$

This is just an extension of the notion introduced by POPRUGENKO [1928]. A typical example of a measure having the Darboux property is the Lebesgue measure on the Borel subsets of $[0,1]$. By Corollary 10.4, finite T_∞-measures on generated tribes have the Darboux property.

12.3 Proposition. *Let $\boldsymbol{m} = (\boldsymbol{m}_1, \boldsymbol{m}_2, \ldots, \boldsymbol{m}_n)$ be an n-vector T_∞-measure on $\mathscr{T}$. Then $\boldsymbol{m}$ has the Darboux property if and only if $\boldsymbol{m}$ is semiconvex in the sense that* *for each $F \in \mathcal{T}$ there exists a fuzzy set $W \in \mathcal{T}(F)$ such that*

$$\boldsymbol{m}(W) = \frac{1}{2} \cdot \boldsymbol{m}(F).$$

PROOF: Assume that $\boldsymbol{m}$ has the Darboux property. Then for each $F \in \mathcal{T}$ and for $E = \emptyset$ there exists a family $(W_r)_{r\in[0,1]} \subseteq \mathcal{T}$ such that (12.1) and (12.2) hold. In particular, $W = W_{1/2} \in \mathcal{T}(F)$ and $\boldsymbol{m}(W) = \frac{1}{2} \cdot \boldsymbol{m}(F)$, and the semiconvexity of $\boldsymbol{m}$ follows. Conversely, let $\boldsymbol{m}$ be semiconvex, consider $E, F \in \mathcal{T}$ such that $E \leq F$ and denote $W_0 = V_0 = E$ and $W_1 = F$. Then there exists a fuzzy set $U_1 \in \mathcal{T}(W_1 \ominus V_0)$ such that $\boldsymbol{m}(U_1) = \frac{1}{2} \cdot \boldsymbol{m}(W_1 \ominus V_0)$. Denoting $V_1 = V_0 \, \mathbf{S}_\infty \, U_1$ and using the T_∞-disjointness of V_0 and U_1, we obtain

$$\boldsymbol{m}(V_1) = \boldsymbol{m}(V_0) + \boldsymbol{m}(U_1) = \frac{1}{2} \cdot (\boldsymbol{m}(W_1) + \boldsymbol{m}(W_0)).$$

By the semiconvexity again, there is a fuzzy set $U_2 \in \mathcal{T}(W_1 \ominus V_1)$ such that

$$\boldsymbol{m}(U_2) = \frac{1}{2} \cdot \boldsymbol{m}(W_1 \ominus V_1).$$

Denote $V_2 = V_1 \, \mathbf{S}_\infty \, U_2$, then we get

$$\boldsymbol{m}(V_2) = \boldsymbol{m}(V_1) + \boldsymbol{m}(U_2) = \frac{3}{4} \cdot \boldsymbol{m}(W_1) + \frac{1}{4} \cdot \boldsymbol{m}(W_0).$$

Continuing in this way we can generate sequences $(U_n)_{n\in\mathbb{N}}$ and $(V_n)_{n\in\mathbb{N}}$ in $\mathcal{T}$ such that $(U_n)_{n\in\mathbb{N}}$ is T_∞-disjoint, $(V_n)_{n\in\mathbb{N}}$ is nondecreasing, and for each $n \in \mathbb{N}$ we have

$$\begin{aligned} \boldsymbol{m}(V_n) &= \frac{1}{2^n} \cdot \boldsymbol{m}(W_0) + (1 - \frac{1}{2^n}) \cdot \boldsymbol{m}(W_1), \\ \boldsymbol{m}(U_n) &= \frac{1}{2} \cdot \boldsymbol{m}(W_1 \ominus V_{n-1}) = \frac{1}{2} \cdot (\boldsymbol{m}(W_1) - \boldsymbol{m}(V_{n-1})). \end{aligned}$$

Choose an arbitrary $r \in \,]0,1[$ and let

$$r = \sum_{i=1}^{\infty} \frac{\rho_i}{2^i}$$

be its (unique) nonterminal binary representation (i.e., the binary representation with $\rho_i \in \{0,1\}$ and $\rho_i \neq 0$ for infinitely many $i \in \mathbb{N}$). For each $k \in \mathbb{N}$ define

$$r_k = \sum_{i=1}^{k} \frac{\rho_i}{2^i}$$

and

$$Z_{r_k} = W_0 \, \mathbf{S}_\infty \, (\mathop{\mathbf{S}_\infty}_{i=1}^{k} \rho_i \cdot U_i).$$

Clearly, $Z_{r_k} \in \mathscr{T}(W_1)$ and

$$\boldsymbol{m}(Z_{r_k}) = \boldsymbol{m}(W_0) + \sum_{i=1}^{k} \rho_i \cdot \boldsymbol{m}(U_i) = r_k \cdot \boldsymbol{m}(W_1) + (1 - r_k) \cdot \boldsymbol{m}(W_0).$$

The sequence $(Z_{r_k})_{k \in \mathbb{N}}$ is nondecreasing. Hence, the limit $W_r = \lim_{k\to\infty} Z_{r_k}$ exists, belongs to $\mathscr{T}(W_1)$ and satisfies

$$\boldsymbol{m}(W_r) = \lim_{k\to\infty} \boldsymbol{m}(Z_{r_k}) = r \cdot \boldsymbol{m}(W_1) + (1 - r) \cdot \boldsymbol{m}(W_0).$$

Obviously for each $r \in [0,1]$ we have $W_0 \leq W_r \leq W_1$, and thus the proposition is proved. □

12.4 Proposition. *An n-vector T_∞-measure $\boldsymbol{m} = (\boldsymbol{m}_1, \boldsymbol{m}_2, \ldots, \boldsymbol{m}_n)$ on $\mathscr{T}$ has the Darboux property if and only if for each pair (E, F) of fuzzy sets in $\mathscr{T}$ with $E \leq F$ there exists a family of fuzzy sets $(W_r)_{r\in[0,1]} \subseteq \mathscr{T}$ such that (12.1), (12.2) and*

$$W_r \leq W_s \quad \textit{whenever} \quad 0 \leq r < s \leq 1 \tag{12.3}$$

are satisfied.

PROOF: Sufficiency trivially holds. Assume that $\boldsymbol{m}$ has the Darboux property and start with $W_0 = E$ and $W_1 = F$. Then there exists $Z_{1/2} \in \mathscr{T}(W_1 \ominus W_0)$ such that $\boldsymbol{m}(Z_{1/2}) = \frac{1}{2} \cdot \boldsymbol{m}(W_1 \ominus W_0)$ (cf. Proposition 12.3). For $W_{1/2} = W_0 \, \mathbf{S}_\infty \, Z_{1/2}$ we have

$$\boldsymbol{m}(W_{1/2}) = \frac{1}{2} \cdot \boldsymbol{m}(W_1) + \frac{1}{2} \cdot \boldsymbol{m}(W_0).$$

Using again the semiconvexity of $\boldsymbol{m}$, we obtain a fuzzy set $Z_{1/4} \in \mathscr{T}(W_{1/2} \ominus W_0)$ and a fuzzy set $Z_{3/4} \in \mathscr{T}(W_1 \ominus W_{1/2})$ such that

$$\begin{aligned} \boldsymbol{m}(Z_{1/4}) &= \frac{1}{2} \cdot \boldsymbol{m}(W_{1/2} \ominus W), \\ \boldsymbol{m}(W_{3/4}) &= \frac{1}{2} \cdot \boldsymbol{m}(W_1 \ominus W_{1/2}). \end{aligned}$$

Denote $W_{1/4} = W_0 \, \mathbf{S}_\infty \, Z_{1/4}$ and $W_{3/4} = W_{1/2} \, \mathbf{S}_\infty \, Z_{3/4}$. It is clear that

$$W_0 \leq W_{1/4} \leq W_{1/2} \leq W_{3/4} \leq W_1$$

and

$$\begin{aligned} \boldsymbol{m}(W_{1/4}) &= \frac{1}{4} \cdot \boldsymbol{m}(W_1) + \frac{3}{4} \cdot \boldsymbol{m}(W_0), \\ \boldsymbol{m}(W_{3/4}) &= \frac{3}{4} \cdot \boldsymbol{m}(W_1) + \frac{1}{4} \cdot \boldsymbol{m}(W_0). \end{aligned}$$

In this way we obtain for each $n \in \mathbb{N}$ a finite family $(W_{k/2^n})_{k\in\{1,2,\ldots,2^n-1\}}$ of fuzzy sets in $\mathscr{T}$ such that

$$W_0 \le W_{1/2^n} \le W_{2/2^n} \le \ldots \le W_{(2^n-1)/2^n} \le W_1$$

and

$$\boldsymbol{m}(W_{k/2^n}) = \frac{k}{2^n} \cdot \boldsymbol{m}(W_1) + (1 - \frac{k}{2^n}) \cdot \boldsymbol{m}(W_0)$$

for all $k \in \{0, 1, \ldots, 2^n\}$. Each $r \in]0,1[$ can be written as $r = \lim_{n\to\infty} r_k$ with r_k as above. The fuzzy set $W_r = \lim_{k\to\infty} W_{r_k}$ is contained in $\mathscr{T}$ and satisfies (12.2). A simple computation shows that the family $(W_r)_{r\in[0,1]}$ also fulfills (12.1) and (12.3). □

12.5 Corollary. *An n-vector T_∞-measure $\boldsymbol{m} = (\boldsymbol{m}_1, \ldots, \boldsymbol{m}_n)$ on $\mathscr{T}$ has the Darboux property if and only if for each $E \in \mathscr{T}$ there exists a family of fuzzy sets $(W_r)_{r\in[0,1]}$ in $\mathscr{T}$ such that the condition (12.3) holds and for each $r \in [0,1]$*

$$\boldsymbol{m}(W_r) = r \cdot \boldsymbol{m}(E) \tag{12.4}$$

12.6 Remark. Halmos [1948] called n-vector measures $\boldsymbol{m}$ (defined on a σ-algebra $\mathscr{A}$ of crisp subsets of X) *convex measures* if the following property holds:

For each $E \in \mathscr{A}$ the set $\mathscr{K}(\boldsymbol{m}, E)$ of all $(\mathscr{A} \cap E)$-measurable functions $g : E \longrightarrow [0,1[$, such that $\boldsymbol{m}(\{x \in E \mid g(x) < r\}) = r \cdot \boldsymbol{m}(E)$ for all $r \in [0,1]$, is convex. Here $\mathscr{A} \cap E$ denotes the restriction of the σ-algebra $\mathscr{A}$ to E, i.e.,

$$\mathscr{A} \cap E = \{A \cap E \mid A \in \mathscr{A}\}.$$

Combining Lemma 4 from Halmos [1948] and Corollary 12.5, it follows that each convex measure has the Darboux property. Since the convexity of an n-vector measure $\boldsymbol{m}$ implies its semiconvexity (cf. [Halmos 1948, Lemma 2]), an n-vector measure $\boldsymbol{m}$ (defined on a σ-algebra of crisp subsets of X) is convex if and only if it has the Darboux property.

12.7 Now we are able to state the main result of this section, namely, a convexity theorem of Liapounoff type for vector T_∞-measures with Darboux property:

Theorem. *If $\boldsymbol{m} = (\boldsymbol{m}_1, \ldots, \boldsymbol{m}_n)$ is an n-vector T_∞-measure on $\mathscr{T}$ with Darboux property, then its range $\mathcal{R}(\boldsymbol{m})$ is convex.*

Proof: It suffices to prove that for each pair (E, F) of fuzzy sets in $\mathscr{T}$ and for all $r \in [0,1]$ there exists a fuzzy set $G \in \mathscr{T}$ such that

$$\boldsymbol{m}(G) = r \cdot \boldsymbol{m}(E) + (1 - r) \cdot \boldsymbol{m}(F). \tag{12.5}$$

First assume that $E\,\mathbf{T}_\infty\, F = \emptyset$. Then there exist $E_r \in \mathscr{T}(E)$ and $F_{1-r} \in \mathscr{T}(F)$ with $\boldsymbol{m}(E_r) = r \cdot \boldsymbol{m}(E)$ and $\boldsymbol{m}(F_{1-r}) = (1-r)\cdot \boldsymbol{m}(F)$ (cf. Corollary 12.5). Denote $G = E_r\,\mathbf{S}_\infty\, F_{1-r}$. Since $E_r\,\mathbf{T}_\infty\, F_{1-r} = \emptyset$, we have proven (12.5) in this case.

Consider the case that E and F are not T_∞-disjoint. Then, however, $A = E \ominus (E\,\mathbf{T}_\infty\, F)$, $B = F \ominus (E\,\mathbf{T}_\infty\, F)$ and $C = E\,\mathbf{T}_\infty\, F$ are T_∞-disjoint. Hence, there exist $A_r \in \mathscr{T}(A)$ and $B_{1-r} \in \mathscr{T}(B)$ such that $A_r\,\mathbf{T}_\infty\, B_{1-r} = \emptyset$ and

$$\boldsymbol{m}(A_r\,\mathbf{S}_\infty\, B_{1-r}) = r \cdot \boldsymbol{m}(A) + (1-r)\cdot \boldsymbol{m}(B).$$

Denote $G = (A_r\,\mathbf{S}_\infty\, B_{1-r})\,\mathbf{S}_\infty\, C$. Then

$$\boldsymbol{m}(G) = r \cdot \boldsymbol{m}(A) + (1-r)\cdot \boldsymbol{m}(B) + \boldsymbol{m}(C) \tag{12.6}$$

since A_r, B_{1-r} and C are T_∞-disjoint. From

$$\begin{aligned} \boldsymbol{m}(A) &= \boldsymbol{m}(E) - \boldsymbol{m}(C), \\ \boldsymbol{m}(B) &= \boldsymbol{m}(F) - \boldsymbol{m}(C) \end{aligned}$$

it follows that (12.5) holds in this case too (cf. (12.6)). □

12.8 In the special case of the finite T_∞-measures Theorem 12.7 can be restated as follows:

Corollary. *If $\boldsymbol{m}$ is a finite T_∞-measure on $\mathscr{T}$ having the Darboux property and if (P,Q) is a Hahn decomposition of X with respect to the σ-additive measure $\overset{\vee}{\boldsymbol{m}}$ on $\mathscr{T}^\vee$, then for each $E \in \mathscr{T}$*

$$\{\boldsymbol{m}(W) \mid W \in \mathscr{T}(E)\} = [\boldsymbol{m}(Q\cdot E), \boldsymbol{m}(P \cdot E)].$$

In particular, we have

$$\mathcal{R}(\boldsymbol{m}) = [\boldsymbol{m}(Q), \boldsymbol{m}(P)].$$

PROOF: For $E \in \mathscr{T}$ and $W \in \mathscr{T}(E)$ we get

$$\boldsymbol{m}(E\cdot Q) = -\boldsymbol{m}^-(E) \le -\boldsymbol{m}^-(W) \le \boldsymbol{m}(W) \le \boldsymbol{m}^+(W) \le \boldsymbol{m}^+(E) = \boldsymbol{m}(P\cdot E).$$

If $r \in [\boldsymbol{m}(E\cdot Q), \boldsymbol{m}(E\cdot P)]$ then there is some $s \in [0,1]$

$$r = \begin{cases} s\cdot \boldsymbol{m}(E\cdot P) & \text{if } r \ge 0, \\ s \cdot \boldsymbol{m}(E\cdot Q) & \text{if } r < 0. \end{cases}$$

Hence, in both cases there exists a fuzzy set $W \in \mathscr{T}(E)$ such that $\boldsymbol{m}(W) = r$ (cf. Corollary 12.5). □

12.9 Remark. Taking into account Corollary 12.8, a finite T_∞-measure $\boldsymbol{m}$ on $\mathscr{T}$ has the Darboux property if and only if for each $E \in \mathscr{T}$ the set $\{\boldsymbol{m}(W) \mid W \in \mathscr{T}(E)\}$ is convex, i.e., if and only if this set is an interval. Generally speaking, if $\boldsymbol{m} = (\boldsymbol{m}_1, \ldots, \boldsymbol{m}_n)$ is an n-vector T_∞-measure on $\mathscr{T}$ with $n \geq 2$ and if for each $E \in \mathscr{T}$ the set

$$\mathcal{R}_E(\boldsymbol{m}) = \{\boldsymbol{m}(W) \mid W \in \mathscr{T}(E)\}$$

is convex, then $\boldsymbol{m}$ has the Darboux property (cf. Proposition 12.3). However, we do not know whether the converse also holds. Clearly, for $E = X$ the set $\mathcal{R}_E(\boldsymbol{m})$ is convex by virtue of Theorem 12.7. But the proof of Theorem 12.7 can not be applied for $E \neq X$ since, in general, we do not have $A\, \mathbf{S}_\infty\, B \in \mathscr{T}(E)$ for each pair (A, B) in $\mathscr{T}(E)$. Such an implication, however, is substantially involved in the proof.

13 Nonatomic T_∞-Measures

13.1 In the preceding section we have shown that the range $\mathcal{R}(\boldsymbol{m})$ of an n-vector T_∞-measure $\boldsymbol{m} = (\boldsymbol{m}_1, \boldsymbol{m}_2, \ldots, \boldsymbol{m}_n)$ with Darboux property is convex. Under which conditions for the components $\boldsymbol{m}_i$ of $\boldsymbol{m}$ can we deduce the Darboux property for $\boldsymbol{m}$? Note that if $\boldsymbol{m}$ has the Darboux property, each of its components $\boldsymbol{m}_1, \boldsymbol{m}_2, \ldots, \boldsymbol{m}_n$ necessarily has the Darboux property. In this section we show that the converse is also true. This is one of Butnariu's results [BUTNARIU 1983b, 1985], and it involves ideas contained in HALMOS' [1948] proof of the Liapounoff theorem.

13.2 Let $\mathscr{T}$ be a T_∞-tribe. All T_∞-measures considered in this section are assumed to be defined on $\mathscr{T}$. If $\boldsymbol{m} = (\boldsymbol{m}_1, \boldsymbol{m}_2, \ldots, \boldsymbol{m}_n)$ is an n-vector T_∞-measure, then define the finite T_∞-measure $\boldsymbol{m}^t : \mathscr{T} \longrightarrow \mathbb{R}$ by

$$\boldsymbol{m}^t = \boldsymbol{m}_1^* + \boldsymbol{m}_2^* + \cdots + \boldsymbol{m}_n^*.$$

Definition. Let $\boldsymbol{m} = (\boldsymbol{m}_1, \boldsymbol{m}_2, \ldots, \boldsymbol{m}_n)$ be an n-vector T_∞-measure on $\mathscr{T}$.

(i) A T_∞-measure $\boldsymbol{n}$ on $\mathscr{T}$ is called ***m*-*absolutely continuous*** if it is $\boldsymbol{m}^t$-absolutely continuous, i.e., if $\boldsymbol{n} \ll \boldsymbol{m}^t$.

(ii) A T_∞-measure $\boldsymbol{n}$ on $\mathscr{T}$ is called *atomic* if there exists a fuzzy set $A \in \mathscr{T}$ with $\boldsymbol{n}(A) \neq \emptyset$ such that for each $B \in \mathscr{T}(A)$ we have either $\boldsymbol{n}(B) = 0$ or $\boldsymbol{n}(B) = \boldsymbol{n}(A)$.

(iii) T_∞-measure $\boldsymbol{n}$ on $\mathscr{T}$ is called *nonatomic* if it is not atomic, i.e., if for each $A \in \mathscr{T}$ with $\boldsymbol{n}(A) \neq \emptyset$ there exists a $B \in \mathscr{T}(A)$ such that $\boldsymbol{n}(B) \notin \{0, \boldsymbol{n}(A)\}$.

(iv) The n-vector T_∞-measure $\boldsymbol{m}$ is called *nonatomic* if each of its components $\boldsymbol{m}_1, \boldsymbol{m}_2, \ldots, \boldsymbol{m}_n$ is nonatomic.

13.3 Theorem. *Let $\boldsymbol{m} = (\boldsymbol{m}_1, \boldsymbol{m}_2, \ldots, \boldsymbol{m}_n)$ be an n-vector T_∞-measure on $\mathscr{T}$. The following are equivalent:*

(i) $\boldsymbol{m}$ has the Darboux property;

(ii) for each $i \in \{1, 2, \ldots, n\}$ the component $\boldsymbol{m}_i$ has the Darboux property;

(iii) $\boldsymbol{m}$ is nonatomic.

PROOF: The implications (i) $\Rightarrow$ (ii) and (ii) $\Rightarrow$ (iii) are trivial. It remains to show that (iii) $\Rightarrow$ (ii) and (ii) $\Rightarrow$ (i). The proof of these implications will be done in the following sequence of lemmas. □

13.4 Lemma. *Assume that $E, F \in \mathscr{T}$ with $E\, \mathbf{T}_\infty\, F = \emptyset$, and let*

$$\boldsymbol{w} = (\boldsymbol{w}_1, \boldsymbol{w}_2, \ldots, \boldsymbol{w}_n)$$

be a nonnegative n-vector T_∞-measure on $\mathscr{T}$ with Darboux property and $\boldsymbol{n}$ a finite T_∞-measure on $\mathscr{T}$ such that $\boldsymbol{n} \ll \boldsymbol{w}$. Then there exists a family of fuzzy sets $(H_r)_{r\in[0,1]}$ in $\mathscr{T}$ such that for each $r \in [0,1]$

$$\boldsymbol{w}(H_r) = r \cdot \boldsymbol{w}(E) + (1 - r) \cdot \boldsymbol{w}(F), \tag{13.1}$$

$$H_0 = F, \qquad H_1 = E. \tag{13.2}$$

Moreover, for the function $f_{\boldsymbol{n}} : [0,1] \longrightarrow \mathbb{R}$ given by $f_{\boldsymbol{n}}(r) = \boldsymbol{n}(H_r)$ we have

$$f_{\boldsymbol{n}} \text{ is continuous.} \tag{13.3}$$

PROOF: If $\boldsymbol{w}(E) = 0$ or $\boldsymbol{w}(F) = 0$, then such a family $(H_r)_{r\in[0,1]}$ exists by Corollary 12.5. Assume that $\boldsymbol{w}(E) \neq 0$ and $\boldsymbol{w}(F) \neq 0$. Since $\boldsymbol{w}$ has the Darboux property, by Proposition 12.4 there are two families $(W_r)_{r\in[0,1]}$ and $(Z_r)_{r\in[0,1]}$ in $\mathscr{T}$ such that for $r, s \in [0,1]$ with $r < s$ we have

$$W_0 = Z_0 = \emptyset, \qquad W_1 = E, \qquad Z_1 = F, \tag{13.4}$$

$$W_r \leq W_s, \qquad Z_r \leq Z_s, \tag{13.5}$$

$$\boldsymbol{w}(W_r) = r \cdot \boldsymbol{w}(E), \qquad \boldsymbol{w}(Z_r) = r \cdot \boldsymbol{w}(F). \tag{13.6}$$

Since $E\,\mathbf{T}_\infty\,F = \emptyset$ it follows that $W_r\,\mathbf{T}_\infty\,T_{1-r} = \emptyset$ for all $r \in [0,1]$. Denote $H_r = W_r\,\mathbf{S}_\infty\,Z_{1-r}$. Then (13.1) and (13.2) are clearly satisfied. Choose $s, r \in [0,1]$ with $s > r$. Then

$$\begin{aligned} |f_{\boldsymbol{n}}(r) - f_{\boldsymbol{n}}(s)| &= |\boldsymbol{n}(Z_{1-r} \ominus Z_{1-s}) - \boldsymbol{n}(W_s \ominus W_r)| \\ &\leq |\boldsymbol{n}(W_s \ominus W_r)| + |\boldsymbol{n}(Z_{1-r} \ominus Z_{1-s})|. \end{aligned} \tag{13.7}$$

For a given $\varepsilon > 0$ there exists a $\delta_\varepsilon > 0$ such that for each $A \in \mathscr{T}$ we have

$$|\boldsymbol{n}(A)| < \frac{\varepsilon}{2} \quad \text{whenever} \quad \boldsymbol{w}^t(A) < \delta_\varepsilon \tag{13.8}$$

because of $\boldsymbol{n} \ll \boldsymbol{w}^t$. Since $\boldsymbol{w}$ is nonnegative, we get

$$\boldsymbol{w}^t(W_s \ominus W_r) = \boldsymbol{w}^t(W_s) - \boldsymbol{w}^t(W_r) = (s - r) \cdot \boldsymbol{w}^t(E)$$

and

$$\boldsymbol{w}^t(Z_{1-r} \ominus Z_{1-s}) = \boldsymbol{w}^t(Z_{1-r}) - \boldsymbol{w}^t(Z_{1-s}) = (s - r) \cdot \boldsymbol{w}^t(F)$$

because of (13.4) and (13.4). Denote

$$\eta_\varepsilon = \min\left(1, \frac{\delta_\varepsilon}{\boldsymbol{w}^t(E)}, \frac{\delta_\varepsilon}{\boldsymbol{w}^t(F)}\right).$$

If $0 < s - r < \eta_\varepsilon$ then we deduce (cf. (13.8))

$$\begin{aligned} |\boldsymbol{n}(W_s \ominus W_r)| &< \frac{\varepsilon}{2}, \\ |\boldsymbol{n}(Z_{1-r} \ominus Z_{1-s})| &< \frac{\varepsilon}{2}. \end{aligned}$$

Taking into account (13.7) it follows that $|f_{\boldsymbol{n}}(r) - f_{\boldsymbol{n}}(s)| < \varepsilon$ whenever $0 < s - r < \eta(\varepsilon)$. Hence, $f_{\boldsymbol{n}}$ is continuous. □

13.5 Lemma. *If $\boldsymbol{w} = (\boldsymbol{w}_1, \boldsymbol{w}_2, \ldots, \boldsymbol{w}_n)$ is a nonnegative n-vector T_∞-measure on $\mathscr{T}$ which has the Darboux property, then for each finite T_∞-measure $\boldsymbol{n}$ with $\boldsymbol{n} \ll \boldsymbol{w}$ the $(n+1)$-vector T_∞-measure $\boldsymbol{w}' = (\boldsymbol{w}_1, \boldsymbol{w}_2, \ldots, \boldsymbol{w}_n, \boldsymbol{n})$ has the Darboux property.*

PROOF: It is sufficient to show that $\boldsymbol{w}'$ is semiconvex (cf. Proposition 12.3). Suppose that $A \in \mathscr{T}$ satisfies $\boldsymbol{w}(A) \neq 0$. Then there exists an $E \in \mathscr{T}(A)$ with $\boldsymbol{w}(E) = \frac{1}{2} \cdot \boldsymbol{w}(A)$. If $\boldsymbol{n}(E) = \frac{1}{2} \cdot \boldsymbol{n}(A)$, we are done. Assume that $\boldsymbol{n}(E) \neq \frac{1}{2} \cdot \boldsymbol{n}(A)$ and denote $F = A \ominus E$. Clearly, $\frac{1}{2} \cdot \boldsymbol{n}(A)$ is contained in the interval with the boundaries $\boldsymbol{n}(E)$ and $\boldsymbol{n}(F)$. Since E and F satisfy Lemma 13.4, there exists a family $(H_r)_{r\in[0,1]}$ in $\mathscr{T}$ such that (13.1), (13.2) and (13.3) hold. Hence, there is an $r_0 \in [0,1]$ such that

$$\boldsymbol{n}(H_{r_0}) = f_{\boldsymbol{n}}(r_0) = \frac{1}{2} \cdot \boldsymbol{n}(A)$$

because of $f_{\boldsymbol{n}}(0) = \boldsymbol{n}(F)$, $f_{\boldsymbol{n}}(1) = \boldsymbol{n}(E)$ and the continuity of $f_{\boldsymbol{n}}$. For $H = H_{r_0}$ we deduce that $\boldsymbol{n}(H) = \frac{1}{2} \cdot \boldsymbol{n}(A)$ and

$$\boldsymbol{w}(H) = r_0 \cdot \boldsymbol{w}(E) + (1 - r_0) \cdot \boldsymbol{w}(F) = \frac{1}{2} \cdot \boldsymbol{w}(A),$$

i.e., $\boldsymbol{w}'(H) = \frac{1}{2} \cdot \boldsymbol{w}(A)$. □

13.6 Lemma. *If $\boldsymbol{m}$ and $\boldsymbol{n}$ are finite T_∞-measures on $\mathscr{T}$, if $\boldsymbol{n} \ll \boldsymbol{m}$ and if $\boldsymbol{m}$ is nonnegative and has the Darboux property, then $\boldsymbol{n}$ also has the Darboux property. Moreover, $\boldsymbol{m} \pm \boldsymbol{n}$ are finite T_∞-measures with Darboux property.*

PROOF: Indeed, if in Lemma 13.5 we take $n = 1$ and $\boldsymbol{w}_1 = \boldsymbol{m}$, then the 2-vector measure $(\boldsymbol{m}, \boldsymbol{n})$ has the Darboux property, and the result follows. □

13.7 Lemma. *If $\boldsymbol{m} = (\boldsymbol{m}_1, \boldsymbol{m}_2, \ldots, \boldsymbol{m}_n)$ is a nonnegative n-vector T_∞-measure on $\mathscr{T}$ with $\boldsymbol{m}_i \ll \boldsymbol{m}_{i-1}$ for all $i \in \{2, 3, \ldots, n\}$ and if $\boldsymbol{m}_1$ has the Darboux property, then $\boldsymbol{m}$ itself has the Darboux property.*

PROOF: For $n = 1$ the result is trivial. Assume that the result holds for each $k < n$. In order to show its validity for $k = n$ consider the $(n-1)$-vector T_∞-measure $\boldsymbol{w} = (\boldsymbol{m}_1, \boldsymbol{m}_2, \ldots, \boldsymbol{m}_{n-1})$ which has the Darboux property. Since

$$\boldsymbol{m}_n \ll \boldsymbol{m}_{n-1} \ll \boldsymbol{m}_1 + \boldsymbol{m}_2 + \cdots + \boldsymbol{m}_{n-1} = \boldsymbol{w}^t,$$

it follows that $\boldsymbol{m}_n \ll \boldsymbol{w}$. Hence, $(\boldsymbol{w}, \boldsymbol{m}_n) = (\boldsymbol{m}_1, \boldsymbol{m}_2, \ldots, \boldsymbol{m}_n)$ has the Darboux property by Lemma 13.5. □

13.8 Lemma. *If $\boldsymbol{n}$ is a finite nonatomic nonnegative T_∞-measure on $\mathscr{T}$, then for each $A \in \mathscr{T}$ with $\boldsymbol{n}(A) \neq 0$ there exists a sequence $(A_k)_{k \in \mathbb{N}_0}$ in $\mathscr{T}$ such that for all $k \in \mathbb{N}$*

$$A = A_0 \geq A_1 \geq A_2 \geq \ldots \geq A_k, \tag{13.9}$$

$$0 < \boldsymbol{n}(A_k) \leq \frac{1}{2} \cdot \boldsymbol{n}(A_{k-1}), \tag{13.10}$$

$$\lim_{k \to \infty} A_k = \emptyset. \tag{13.11}$$

PROOF: Put $B_0 = A$ and assume that we have already obtained fuzzy sets $B_0 \geq B_1 \geq B_2 \geq \ldots \geq B_k$ in $\mathscr{T}$ such that $0 < \boldsymbol{n}(B_i) \leq \frac{1}{2^i} \cdot \boldsymbol{n}(A)$ and $\boldsymbol{n}(B_i) \leq \frac{1}{2} \cdot \boldsymbol{n}(B_{i-1})$ for $i \in \{1, 2, \ldots, k\}$. Then there exists $B \in \mathscr{T}(B_k)$ with $0 < \boldsymbol{n}(B) < \boldsymbol{n}(B_k)$ because of the nonatomicity of $\boldsymbol{n}$. Define B_{k+1} by

$$B_{k+1} = \begin{cases} B & \text{if } \boldsymbol{n}(B) \leq \frac{1}{2} \cdot \boldsymbol{n}(Bk), \\ B_k \ominus B & \text{otherwise.} \end{cases}$$

Clearly

$$0 < \boldsymbol{n}(B_{k+1}) \le \frac{1}{2} \cdot \boldsymbol{n}(B_k)$$

and $B_{k+1} \le B_k$. Moreover, we have

$$\boldsymbol{n}(B_{k+1}) \le \frac{1}{2^{k+1}} \cdot \boldsymbol{n}(A).$$

The sequence $(B_k)_{k\in\mathbb{N}} \subseteq \mathscr{T}$ obtained in this way is nonincreasing and its limit $W = \lim_{k\to\infty} B_k$ belongs to $\mathscr{T}$ (cf. Theorem 2.7). Since $\boldsymbol{n}(B_k) \le \frac{1}{2^k} \cdot \boldsymbol{n}(A)$ for all $k \in \mathbb{N}$, it follows that $\boldsymbol{n}(W) = 0$ because of the continuity of $\boldsymbol{n}$. If we define $A_k = B_k \ominus W$ for $k \in \mathbb{N}$ and $A_0 = A$, then the sequence $(A_k)_{k\in\mathbb{N}_0}$ satisfies the properties (13.9), (13.10) and (13.11). □

13.9 Lemma. *If $\boldsymbol{n}$ is a nonnegative nonatomic finite T_∞-measure on $\mathscr{T}$, then $\boldsymbol{n}$ has the Darboux property.*

PROOF: We prove that $\boldsymbol{n}$ is semiconvex. For each $A \in \mathscr{T}$ with $\boldsymbol{n}(A) \ne 0$, there exists a sequence $(A_k)_{k\in\mathbb{N}_0} \subseteq \mathscr{T}$ such that (13.9), (13.10) and (13.11) are satisfied. Denote $W_k = A_{k-1} \ominus A_k$ for each $k \in \mathbb{N}$. From (13.9) it follows that $(W_k)_{k\in\mathbb{N}}$ is a T_∞-disjoint sequence in $\mathscr{T}(A)$. For $k \in \mathbb{N}$ and $r \in \{0,1\}$ define the fuzzy set

$$W_{k,r} = \begin{cases} W_k & \text{if } r = 1, \\ \emptyset & \text{if } r = 0. \end{cases}$$

If $t \in [0,1]$ and

$$t = \sum_{k=1}^{\infty} \frac{t_k}{2^k}$$

its unique nonterminal binary representation, denote

$$t * W = \mathop{\mathbf{S}_\infty}_{k=1}^{\infty} W_{k,t_k}.$$

Clearly, $t * W \in \mathscr{T}(A)$ and

$$\boldsymbol{n}(t * W) = \sum_{k=1}^{\infty} t_k \cdot \boldsymbol{n}(W_k).$$

We claim that the function $g : [0,1] \longrightarrow \mathbb{R}$ defined by $g(t) = \boldsymbol{n}(t * W)$ is continuous. Indeed, if $0 \le s < t \le 1$, then there exists a smallest $j \in \mathbb{N}$ such that $t_j = 1$ and $s_j = 0$. Hence,

$$g(t) - g(s) = \boldsymbol{n}(W_j) + \sum_{k=j+1}^{\infty} (t_k - s_k) \cdot \boldsymbol{n}(W_k)$$

$$\geq \ \boldsymbol{n}(W_j) - \sum_{k=j+1}^{\infty} \boldsymbol{n}(W_k).$$

Since the latter sum equals $\boldsymbol{n}(A_j)$, it follows that

$$g(t) - g(s) \geq \boldsymbol{n}(A_{j-1}) - 2 \cdot \boldsymbol{n}(A_j) \geq 0$$

because of (13.10). Now observe that

$$\begin{aligned} g(t) - g(s) \ &\leq \ \sum_{k=1}^{\infty}(t_k - s_k) \cdot \boldsymbol{n}(A_{k-1}) \\ &\leq \ 2 \cdot (t - s) \cdot \boldsymbol{n}(A), \end{aligned}$$

and the continuity of g follows. The function g satisfies $g(0) = 0$ and

$$g(1) = \boldsymbol{n}(\mathbf{S}_{\infty}{}_{k=1}^{\infty} W_k) = \boldsymbol{n}(A).$$

Hence, there exists a $t^\circ \in [0,1]$ such that $g(t^\circ) = \frac{1}{2} \cdot g(1)$, i.e., $\boldsymbol{n}(B) = \frac{1}{2} \cdot \boldsymbol{n}(A)$ and $B \in \mathscr{T}(A)$ for $B = t^\circ * W$, and $\boldsymbol{n}$ is semiconvex. □

13.10 Lemma. *If $\boldsymbol{n}_1, \boldsymbol{n}_2, \ldots, \boldsymbol{n}_n$ are nonnegative finite T_∞-measures on $\mathscr{T}$ having the Darboux property, then $\boldsymbol{n} = \boldsymbol{n}_1 + \boldsymbol{n}_2 + \ldots + \boldsymbol{n}_n$ has the Darboux property.*

PROOF: We prove that $\boldsymbol{n}$ is nonatomic. Assume that $A \in \mathscr{T}$ and $\boldsymbol{n}(A) \neq 0$. Then $\boldsymbol{n}_i(A) \neq 0$ for at least one index i. Since $\boldsymbol{n}_i$ has the Darboux property, there exists a fuzzy set $B \in \mathscr{T}(A)$ such that $\boldsymbol{n}_i(B) = \frac{1}{2} \cdot \boldsymbol{n}_i(A) < \boldsymbol{n}_i(A)$. Hence, $0 < \boldsymbol{n}(B) < \boldsymbol{n}(A)$ since all $\boldsymbol{n}_i$ are monotone (cf. Proposition 6.1). □

13.11 Lemma. *If $\boldsymbol{m} = (\boldsymbol{m}_1, \boldsymbol{m}_2 \ldots, \boldsymbol{m}_n)$ is a nonnegative n-vector T_∞-measure on $\mathscr{T}$ and if each $\boldsymbol{m}_1, \boldsymbol{m}_2, \ldots, \boldsymbol{m}_n$ has the Darboux property, then m itself has the Darboux property.*

PROOF: For $i \in \{1, 2, \ldots, n\}$ denote $\boldsymbol{m}_i^0 = \sum_{k=i}^{n} \boldsymbol{m}_k$. Each $\boldsymbol{m}_i^0$ has the Darboux property by Lemma 13.10. Observe that $\boldsymbol{m}_i^0 \ll \boldsymbol{m}_{i-1}^0$ whenever $i \in \{2, 3, \ldots, n\}$. Thus $\boldsymbol{m}^0 = (\boldsymbol{m}_1^0, \boldsymbol{m}_2^0, \ldots, \boldsymbol{m}_n^0)$ has the Darboux property by Lemma 13.5. Let $f = (f_1, \ldots, f_n)$ be the linear operator $f : \mathbb{R}^n \longrightarrow \mathbb{R}^n$ defined by

$$f_i(y_1, y_2, \ldots, y_n) = \sum_{k=i}^{n} y_k.$$

It is clear that f is one-to-one and onto (its associated determinant is different from zero). Therefore, f^{-1} exists and is linear. From $\boldsymbol{m}^0 = f \circ \boldsymbol{m}$ it follows that $\boldsymbol{m} = f^{-1} \circ \boldsymbol{m}^0$ and, subsequently, that $\boldsymbol{m}$ is semiconvex. □

13.12 Now we are ready to show that (ii) implies (i) in Theorem 13.3.

Lemma. *If $\boldsymbol{m} = (\boldsymbol{m}_1, \boldsymbol{m}_2, \ldots, \boldsymbol{m}_n)$ is an n-vector T_∞-measure on $\mathscr{T}$ and if each component $\boldsymbol{m}_1, \boldsymbol{m}_2, \ldots, \boldsymbol{m}_n$ has the Darboux property, then $\boldsymbol{m}$ itself has the Darboux property.*

PROOF: For each index $i \in \{1, 2, \ldots, n\}$ there exists a Hahn decomposition (P_i, Q_i) of X with respect to $\overset{\vee}{\boldsymbol{m}}_i$ (the restriction of $\boldsymbol{m}_i$ to $\mathscr{T}^\vee$). Denote $E_i^0 = P_i$ and $E_i^1 = Q_i$. For $a = (a_1, a_2, \ldots, a_n) \in \{0,1\}^n$ denote $E(a) = \bigcap_{i=1}^n E_i^{a_i}$. Clearly, $E(a) \in \mathscr{T}$. Define the finite T_∞-measures $\boldsymbol{m}_1^a, \boldsymbol{m}_2^a, \ldots, \boldsymbol{m}_n^a$ by

$$\boldsymbol{m}_i^a(A) = (-1)^{a_i} \cdot \boldsymbol{m}_i(A\,\mathbf{T}_\infty\, E(a)).$$

Each $\boldsymbol{m}_i^a$ is nonnegative since

$$A\,\mathbf{T}_\infty\, E(a) \leq \begin{cases} Q_i & if \quad a_i = 1 \\ P_i & if \quad a_i = 0. \end{cases}$$

We can even show that each $\boldsymbol{m}_i^a$ has the Darboux property: if $A \in \mathscr{T}$ and $\boldsymbol{m}_i^a(A) \neq 0$, then there exists a $B \in \mathscr{T}(A)$ such that

$$\boldsymbol{m}_i^a(B) = (-1)^{a_i} \cdot \boldsymbol{m}_i(B\,\mathbf{T}_\infty\, E(a)) = \frac{1}{2} \cdot (-1)^{a_i} \cdot \boldsymbol{m}_i(A\,\mathbf{T}_\infty\, E(a)) = \frac{1}{2} \cdot \boldsymbol{m}_i^a(A)$$

since $\boldsymbol{m}_i$ is semiconvex and $E(a)$ is a subset of X. Hence, $\boldsymbol{m}^a = (\boldsymbol{m}_1^a, \boldsymbol{m}_2^a, \ldots, \boldsymbol{m}_n^a)$ is a nonnegative n-vector T_∞-measure with Darboux property (cf. Lemma 13.11). Since $\{E(a) \mid a \in \{0,1\}^n\}$ is a partition of X, we have

$$\boldsymbol{m} = \sum_{a \in \{0,1\}^n} (-1)^a \cdot \boldsymbol{m}^a, \tag{13.12}$$

where the n-vector T_∞-measure $(-1)^a \cdot \boldsymbol{m}^a$ is given by

$$(-1)^a \cdot \boldsymbol{m}^a = ((-1)^{a_1} \cdot \boldsymbol{m}_1^a, (-1)^{a_2} \cdot \boldsymbol{m}_2^a, \ldots, (-1)^{a_n} \cdot \boldsymbol{m}_n^a).$$

By the Darboux property of each $\boldsymbol{m}^a$, for each $A \in \mathscr{T}$ and for each $a \in \{0,1\}^n$, there exists an $A(a) \in \mathscr{T}(A)$ such that $\boldsymbol{m}^a(A(a)) = \frac{1}{2} \cdot \boldsymbol{m}^a(A)$. Denote

$$B(a) = A(a)\,\mathbf{T}_\infty\, E(a)$$

and

$$B = \sum_{a \in \{0,1\}^n} B(a).$$

Then we have

$$\begin{aligned} \boldsymbol{m}(B) &= \sum_{a\in\{0,1\}^n} \boldsymbol{m}(B(a)) \\ &= \sum_{a\in\{0,1\}^n} (-1)^a \cdot \boldsymbol{m}^a(A(a)) \\ &= \frac{1}{2}\cdot \sum_{a\in\{0,1\}^n} (-1)^a \cdot \boldsymbol{m}^a(A) \\ &= \frac{1}{2}\cdot \boldsymbol{m}(A) \end{aligned}$$

because of (13.12) and the T_∞-disjointness of the family $(B(a))_{a\in\{0,1\}^n}$. This shows that $\boldsymbol{m}$ is semiconvex. □

13.13 The following lemma verifies implication (iii) ⇒ (ii) of Theorem 13.3, and this will complete the proof of this theorem.

Lemma. *If $\boldsymbol{m} = (\boldsymbol{m}_1, \boldsymbol{m}_2, \ldots, \boldsymbol{m}_n)$ is a nonatomic n-vector T_∞-measure on $\mathscr{T}$ then each component $\boldsymbol{m}_1, \boldsymbol{m}_2, \ldots, \boldsymbol{m}_n$ has the Darboux property.*

PROOF: If $\boldsymbol{m}$ is nonatomic, each of its components $\boldsymbol{m}_i$ is finite and nonatomic. If each $\boldsymbol{m}_i$ is nonnegative, then the result follows from Lemma 13.9. Assume that $\boldsymbol{m}_i$ has also some negative values. If we can show that $\boldsymbol{m}_i^+$ and $\boldsymbol{m}_i^-$ are also nonatomic, then they have the Darboux property (cf. Lemma 13.9) and, consequently, the 2-vector T_∞-measure $(\boldsymbol{m}_i^+, \boldsymbol{m}_i^-)$ (cf. Lemma 13.11) and the T_∞-measure $\boldsymbol{m}_i = \boldsymbol{m}_i^+ - \boldsymbol{m}_i^-$ also have the Darboux property due to $\boldsymbol{m}_i = f(\boldsymbol{m}_i^+, \boldsymbol{m}_i^-)$ and to the fact that the linear function $f : \mathbb{R}^2 \longrightarrow \mathbb{R}$ defined by $f(y_1, y_2) = y_1 - y_2$ preserves semiconvexity. If $A \in \mathscr{T}$ with $\boldsymbol{m}_i^+(A) \neq 0$, then

$$\boldsymbol{m}_i(P_i \,\mathbf{T}_\infty\, A) = \boldsymbol{m}_i^+(A) \neq 0$$

for each Hahn decomposition (P_i, Q_i) of X with respect to $\check{\boldsymbol{m}}_i$ (cf. Corollary 10.4). Since $\boldsymbol{m}_i$ is nonatomic, there exists a fuzzy set $B \in \mathscr{T}(P_i \,\mathbf{T}_\infty\, A)$ such that

$$0 \neq \boldsymbol{m}_i(B) \neq \boldsymbol{m}_i(P_i \,\mathbf{T}_\infty\, A) = \boldsymbol{m}_i^+(A).$$

As a consequence,

$$\boldsymbol{m}_i^+(B) = \boldsymbol{m}_i(P_i \, T_\infty \, B) = \boldsymbol{m}_i(B) \notin \{0, \boldsymbol{m}_i^+(A)\},$$

i.e., $\boldsymbol{m}_i^+$ is nonatomic. The nonatomicity of $\boldsymbol{m}_i^-$ can be seen analogously. □

13.14 Theorem 13.3 together with Corollary 10.3 allows Theorem 12.7 to be restated as follows:

Corollary. *If $\boldsymbol{m} = (\boldsymbol{m}_1, \ldots, \boldsymbol{m}_n)$ is a finite nonatomic n-vector T_∞-measure on $\mathscr{T}$, then its range $\mathcal{R}(\boldsymbol{m})$ is convex and bounded. In the case of $n = 1$, moreover,*

$$\mathcal{R}(\boldsymbol{m}) = [\boldsymbol{m}(Q), \boldsymbol{m}(P)],$$

(P, Q) being a Hahn decomposition of X with respect to the σ-additive measure $\check{\boldsymbol{m}}$.

14 A Liapounoff Type Theorem for T_∞-Measures

14.1 In Section 13 we have shown that the range $\mathcal{R}(\boldsymbol{m})$ of a nonatomic T_∞-measure $\boldsymbol{m}$ on a T_∞-tribe is convex and bounded (cf. Corollary 13.14). In this section we show that $\mathcal{R}(\boldsymbol{m})$ is also compact. The idea of this proof is extracted from HALMOS' [1948] proof the Liapounoff theorem. Throughout this section $\mathscr{T}$ will denote a T_∞-tribe, and all T_∞-measures are assumed to be defined on $\mathscr{T}$.

14.2 Theorem. *If $\boldsymbol{m} = (\boldsymbol{m}_1, \ldots, \boldsymbol{m}_n)$ is a nonatomic n-vector T_∞-measure on $\mathscr{T}$, then its range $\mathcal{R}(\boldsymbol{m})$ is convex and compact.*

PROOF: Convexity and boundedness of $\mathcal{R}(\boldsymbol{m})$ follow from Corollary 13.14. Without loss of generality, we may assume that the affine dimension of $\mathcal{R}(\boldsymbol{m})$ equals n. Otherwise, all the following arguments can be carried over from $\mathbb{R}^n$ to the linear hull of $\mathcal{R}(\boldsymbol{m})$. It remains to show that $\mathcal{R}(\boldsymbol{m})$ is closed. To this end, it is sufficient to show that

$$H \cap \mathrm{Cl}(\mathcal{R}(\boldsymbol{m})) \subseteq \mathcal{R}(\boldsymbol{m}), \tag{14.1}$$

for each supporting hyperplane H of $\mathcal{R}(\boldsymbol{m})$ (cf. [ROCKAFELLAR 1970, Theorem 11.6.2 and Corollary 11.7.1]). Let H be such a supporting hyperplane. Then there exists an $r \in \mathbb{R}$ and a linear function $f : \mathbb{R}^n \longrightarrow \mathbb{R}$ such that

$$H = \{x \in \mathbb{R}^n \mid f(x) = r\}$$

and for all $E \in \mathscr{T}$

$$f(\boldsymbol{m}(E)) \geq r.$$

The proof is carried out in several steps.

Claim 1. For each crisp set $A \in \mathcal{T}^\vee$ the function $\boldsymbol{m}_A : \mathcal{T} \longrightarrow \mathbb{R}$ defined by

$$\boldsymbol{m}_A(E) = \boldsymbol{m}(A^c \cdot E) - \boldsymbol{m}(A \cdot E) \tag{14.2}$$

is a nonatomic n-vector T_∞-measure and

$$\mathcal{R}(\boldsymbol{m}_A) = \mathcal{R}(\boldsymbol{m}) - \boldsymbol{m}(A).$$

Indeed, a simple computation shows that $\boldsymbol{m}_A$ is a n-vector T_∞-measure. For $E \in \mathcal{T}$ we have

$$\boldsymbol{m}_A(E) = \boldsymbol{m}((A^c \cdot E)\, \mathbf{S}_\infty\, (A \cdot E^c)) - \boldsymbol{m}(A),$$

i.e., $\mathcal{R}(\boldsymbol{m}_A) \subseteq \mathcal{R}(\boldsymbol{m}) - \boldsymbol{m}(A)$. Conversely, since $A \in \mathcal{T}^\vee$, we get

$$\boldsymbol{m}(E) - \boldsymbol{m}(A) = \boldsymbol{m}_A((A^c \cdot E)\, \mathbf{S}_\infty\, (A \cdot E^c)),$$

i.e., $\mathcal{R}(\boldsymbol{m}) - \boldsymbol{m}(A) \subseteq \mathcal{R}(\boldsymbol{m}_A)$.

To show the nonatomicity of $\boldsymbol{m}_A$, observe that for $E \in \mathcal{T}$ there exist T_∞-disjoint fuzzy sets $F_1 \in \mathcal{T}(A^c\, \mathbf{T}_\infty\, E)$ and $F_2 \in \mathcal{T}(A\, \mathbf{T}_\infty\, E)$ such that

$$\begin{aligned} \boldsymbol{m}(F_1) &= \frac{1}{2} \cdot \boldsymbol{m}(A^c\, \mathbf{T}_\infty\, E), \\ \boldsymbol{m}(F_2) &= \frac{1}{2} \cdot \boldsymbol{m}(A\, \mathbf{T}_\infty\, E) \end{aligned}$$

since $\boldsymbol{m}$ is nonatomic (see Theorem 13.3 and Propostition 12.3). Therefore,

$$\boldsymbol{m}_A(F_1\, \mathbf{S}_\infty\, F_2) = \boldsymbol{m}_A(F_1) + \boldsymbol{m}_A(F_2) = \boldsymbol{m}(F_1) - \boldsymbol{m}(F_2) = \frac{1}{2} \cdot \boldsymbol{m}_A(E),$$

showing that $\boldsymbol{m}_A$ is nonatomic.

Claim 2. There exists a set $Q \in \mathcal{T}^\vee$ such that $\boldsymbol{m}(Q) \in H \cap \mathcal{R}(\boldsymbol{m})$.

Define the finite T_∞-measure $\boldsymbol{u} : \mathcal{T} \longrightarrow \mathbb{R}$ by $\boldsymbol{u} = f \circ \boldsymbol{m}$. Observe that

$$r = \inf\{\boldsymbol{u}(E) \mid E \in T\},$$

and let (P, Q) be a Hahn decomposition of X with respect to the σ-additive measure $\overset{\vee}{\boldsymbol{u}}$. By Corollary 10.4 (iv), we have $\boldsymbol{u}(Q) = r$, and this proves Claim 2.
Now we are in the position to show (14.1). Observe that $f \circ \boldsymbol{m}_Q(E) \geq 0$ for all $E \in \mathcal{T}$, and define $\boldsymbol{u}_Q = f \circ \boldsymbol{m}_Q$. Let L be the linear subspace $L = \{x \in \mathbb{R}^n \mid f(x) = 0\}$, i.e., $L = H - \boldsymbol{m}(Q)$. By Claim 1, (14.1) is equivalent to

$$L \cap \mathrm{Cl}(\mathcal{R}(\boldsymbol{m}_Q)) \subseteq \mathcal{R}(\boldsymbol{m}_Q). \tag{14.3}$$

Denote $\boldsymbol{w} = \boldsymbol{m}_Q^t$(see Definition 13.2). Clearly, each coordinate of $\boldsymbol{m}_Q$ is $\boldsymbol{w}$-absolute continuous. We prove (14.3) by induction upon the number n of components of the vector T_∞-measure $\boldsymbol{m}$. If $n = 1$ then (14.3) follows from Corollary 13.14. Assuming that (14.3) is true for all $k < n$, we want to prove it for $k = n$. Then, by Corollary 10.4 (ii) and [HALMOS 1974, p. 125], there exists a crisp set $B \in \mathscr{T}^\vee$ such that $\boldsymbol{u}_Q(B) = 0$ and the n-vector T_∞-measure $\boldsymbol{w}_B$ defined by $\boldsymbol{w}_B(E) = \boldsymbol{w}(B \cdot E)$ is $\boldsymbol{u}_Q$-absolutely continuous. Define the n-vector T_∞-measures $\boldsymbol{w}_1$ and $\boldsymbol{w}_2$ by $\boldsymbol{w}_1(E) = \boldsymbol{m}_Q(B \cdot E)$ and $\boldsymbol{w}_2(E) = \boldsymbol{m}_Q(B^c \cdot E)$, respectively. Since for each $E \in \mathscr{T}$ we have

$$f \circ \boldsymbol{w}_1(E) = f \circ \boldsymbol{m}_Q(B \cdot E) = \boldsymbol{u}_Q(B \cdot E) = 0,$$

we also get $\mathcal{R}(\boldsymbol{w}_1) \subseteq L$. Hence, the affine dimension of $\mathcal{R}(\boldsymbol{w}_1)$ is less than or equal to $n - 1$, implying that $\mathcal{R}(\boldsymbol{w}_1)$ is compact due to the nonatomicity of $\boldsymbol{w}_1$ and the inductive assumption. Let x be a point in $L \cap \mathrm{Cl}(\mathcal{R}(\boldsymbol{m}_Q))$. Then there exists a sequence $(E_l)_{l\in\mathbb{N}} \subseteq \mathscr{T}$ such that $\lim_{l\to\infty} \boldsymbol{m}_Q(E_l) = x$, and therefore,

$$0 = f(x) = f(\lim_{l\to\infty} \boldsymbol{m}_Q(E_l)) = \lim_{l\to\infty} \boldsymbol{u}_Q(E_l).$$

By the $\boldsymbol{u}_Q$-absolute continuity of $\boldsymbol{w}_2$ we get $\lim_{l\to\infty} \boldsymbol{w}_2(E_l) = 0$. Note that $\boldsymbol{m}_Q = \boldsymbol{w}_1 + \boldsymbol{w}_2$ which implies

$$\lim_{l\to\infty} \boldsymbol{m}_Q(E_l) = \lim_{l\to\infty} \boldsymbol{w}_1(E_l) = x,$$

or, in other words,

$$x \in \mathrm{Cl}(\mathcal{R}(\boldsymbol{w}_1)) = \mathcal{R}(\boldsymbol{w}_1) \subseteq \mathcal{R}(\boldsymbol{m}_Q).$$

Hence, (14.3) holds. □

14.3 Remark.

(i) Theorem 14.2 obviously extends the Liapounoff Theorem ([LIAPOUNOFF 1940]).

(ii) Observing that T_∞-measures on generated tribes are necessarily nonatomic (cf. Corollary 10.4 (ii)), Theorem 14.2 contains the theorems by DVORETZKI et al. [1951] stating that *the range of a vector integral is convex and compact* as a special case.

Chapter IV

Games with Fuzzy Coalitions

15 Functions of Bounded Variation

15.1 In this chapter the set X is interpreted as a set of *players* who participate in interactive activities by forming *fuzzy coalitions*, i.e., coalitions which they may belong to in various degrees. Such a coalition will be represented by a fuzzy subset of X. If A is a fuzzy subset of X representing a fuzzy coalition, then we interprete the membership degree of a player $x \in X$ to A as the share of the individual worth of the player x which is *invested* as his participation in the fuzzy coalition A. In a specific social environment the class of fuzzy coalitions which can be effectively constituted by the players is determined by exogeneous rules. We always assume that this class is a T_∞-tribe $\mathscr{T}$ which is given a priori, and the term *fuzzy coalition* is used to determine an element of $\mathscr{T}$. A *game on* $\mathscr{T}$ or, simply, a *game* is a function $\boldsymbol{v} : \mathscr{T} \longrightarrow \mathbb{R}$ with $\boldsymbol{v}(\emptyset) = 0$. The class of all games on $\mathscr{T}$ is denoted $GAMES_{\mathscr{T}}$ or, simply, $GAMES$, when there is no possibility of confusion. $GAMES_{\mathscr{T}}$ is an algebra with the operations naturally induced from $\mathbb{R}$.

Games with fuzzy coalitions, without using this specific term, were studied in the past for various reasons. AUMANN & SHAPLEY [1974] called such games *ideal set functions*, and they used them to establish the relationship between the *value on pNA* and the cores of a class of market games. The term *fuzzy coalition* (*coalition flou*) was introduced by AUBIN [1974 a, b]. AUBIN [1974 a, b, 1979] studied a class of games with fuzzy coalitions and proved the existence of a *partitional value* operator on this class. MERTENS [1980, 1988] employed games with fuzzy coalitions as technical tools in his attempts to extend the *Aumann–Shapley value* on *pNA* to larger spaces of games.

The aim of this and the following chapter is to study games with fuzzy coalitions. Our main purpose is to identify large classes of games with fuzzy coalitions on which an *Aumann–Shapley value* (see Definition 18.2) does exist. The reason for considering this problem is threefold. On the one hand, the Aumann–Shapley value represents a model of *fair* sharing of the outcomes of cooperative behavior in economic games and, therefore, the existence of Aumann–Shapley values is of intrinsic interest in many game theoretical models of cooperative social behavior. On the other hand, it was shown by AUMANN & PERLES [1965] and AUMANN & SHAPLEY [1974] that the problem of finding transferable utility competitive equilibria in large classes of market games is strongly related to the problem of finding values for these games since the Aumann–Shapley values implicitly define the *competitive payoff densities* associated with the transferable utility competitive equilibria (see [AUMANN & SHAPLEY 1974, Chapter VI]). Finally, MERTENS [1980] observed that it is sometimes possible to obtain values on various spaces of games with *crisp* coalitions by *projecting* values existing on specific classes of games with fuzzy coalitions onto the space of games with crisp coalitions.

Our approach to the problem of the existence of Aumann–Shapley values for games with fuzzy coalitions consists, in principle, of extending the Aumann–Shapley value existing on a space we call $pFNA_{\mathscr{T}}$ to spaces which are much larger. The space $pFNA_{\mathscr{T}}$ contains all the games which can be *approximated* by linear combinations of positive integer powers of nonatomic T_∞-measures on $\mathscr{T}$. The main result in this chapter will be Theorem 18.4 showing that there exists an Aumann–Shapley value on $pFNA_{\mathscr{T}}$ whenever $\mathscr{T}$ is a T_∞-tribe (generated or not).

This is a true generalization of the existence part of Theorem B in [AUMANN & SHAPLEY 1974]; the extent of the generalization becomes apparent in Section 19. The proof of Theorem 18.4 follows the main ideas of the proof of the classical theorem, but there are some substantial, conceptual and technical differences due to the greater generality of our result. These differences, together with the fact that some constructive aspects of the proof of Theorem 18.4 are substantially involved in the further extension of the value on $pFNA_{\mathscr{T}}$ to larger subspaces of $GAMES_{\mathscr{T}}$, are the reasons for presenting here the proof of Theorem 18.4 in full length. This is done in spite of the fact that some technical details are well-known, mainly from AUMANN & SHAPLEY [1974, Chapter I].

15.2 We start our considerations recalling the following notions:

Definition. Let $\mathscr{T}$ be a T_∞-tribe on X.

(i) A finite family of fuzzy coalitions $\mathscr{B} = (A_0, A_1, \ldots, A_n) \in \mathscr{T}^n$ is called a *chain* if

$$\emptyset = A_0 \leq A_1 \leq A_2 \leq \ldots \leq A_n = X. \tag{15.1}$$

A pair (A_i, A_{i+1}) of elements of the chain $\mathscr{B}$ is called a *link*. A family of links of $\mathscr{B}$ is called a *subchain* of $\mathscr{B}$.

(ii) If $\boldsymbol{v}$ is a game and $\mathscr{C}$ is a subchain of $\mathscr{B}$, then

$$\|\boldsymbol{v}\|_{\mathscr{C}} = \sum_{(A_i, A_{i+1}) \in \mathscr{C}} |\boldsymbol{v}(A_{i+1}) - \boldsymbol{v}(A_i)| \tag{15.2}$$

is called the *variation of $\boldsymbol{v}$ with respect to $\mathscr{C}$*. If $\mathscr{C}$ contains all the links of $\mathscr{B}$, we write $\|\boldsymbol{v}\|_{\mathscr{B}}$ instead of $\|\boldsymbol{v}\|_{\mathscr{C}}$ and call it the *variation of $\boldsymbol{v}$ with respect to $\mathscr{B}$.*

15.3 We now introduce the following notion which is similar to AUMANN & SHAPLEY's [1974] definitions of the spaces BV and IBV.

Definition. Let $\boldsymbol{v}$ be a game on a T_∞-tribe $\mathscr{T}$. The *total variation* of $\boldsymbol{v}$ is defined by

$$\|\boldsymbol{v}\| = \sup\{\|\boldsymbol{v}\|_{\mathscr{B}} \mid \mathscr{B} \text{ chain of fuzzy coalitions}\}. \tag{15.3}$$

The *game $\boldsymbol{v}$* is said to be *of bounded variation* if $\|\boldsymbol{v}\| < \infty$. The family of games on $\mathscr{T}$ which are of bounded variation is denoted $FBV_{\mathscr{T}}$ or, simply, FBV if there is no possibility of confusion.

15.4 Obviously, $FBV_{\mathscr{T}}$ contains all monotone games on $\mathscr{T}$, i.e., all games such that $\boldsymbol{v}(A) \leq \boldsymbol{v}(B)$ whenever $A \leq B$. Moreover, if $\boldsymbol{v}$ is a monotone game, then we have $\|\boldsymbol{v}\| = \boldsymbol{v}(X)$. Standard reasoning shows that $FBV_{\mathscr{T}}$ is a normed linear space, with the total variation as its norm.

Note that, if the T_∞-tribe $\mathscr{T}$ is a σ-algebra of subsets of X, then $FBV_{\mathscr{T}}$ is exactly the space BV discussed by AUMANN & SHAPLEY [1974]. If $\mathscr{T}$ is a generated T_∞-tribe, then $FBV_{\mathscr{T}}$ equals the space IBV introduced by AUMANN & SHAPLEY [1974]. The nature of the T_∞-tribe $\mathscr{T}$ substantially affects the topological and geometric structure of the space $FBV_{\mathscr{T}}$.

The relation between the spaces $FBV_{\mathscr{T}^\vee}$, $FBV_{\mathscr{T}}$ and $FBV_{\mathscr{T}^\diamond}$ is quite complicated, in spite of the fact that $\mathscr{T}^\vee \subseteq \mathscr{T} \subseteq \mathscr{T}^\diamond$ (cf. Theorem 3.2). This will be illustrated more thoroughly later on, starting in Section 19.

15.5 In this chapter, if not specified otherwise, $\mathscr{T}$ always denotes a fixed T_∞-tribe (generated or not). The following standard characterization of games of bounded variation, in principle given by AUMANN & SHAPLEY [1974], will be useful:

Lemma. *A game $\boldsymbol{v}$ is of bounded variation if and only if there exists a pair $(\boldsymbol{u}, \boldsymbol{w})$ of monotone games such that $\boldsymbol{v} = \boldsymbol{u} - \boldsymbol{w}$.*

PROOF: Assume that $\boldsymbol{v}$ is of bounded variation on $\mathscr{T}$. Define the function $\boldsymbol{u} : \mathscr{T} \longrightarrow \mathbb{R}$ by

$$\boldsymbol{u}(A) = \sup_{\mathscr{B}} \left\{ \sum_{i=1}^{k} \max[\boldsymbol{v}(A_i) - \boldsymbol{v}(A_{i-1}), 0] \right\}, \tag{15.4}$$

where the supremum is taken over all chains of fuzzy coalitions $\mathscr{B} = (A_0, A_1, \ldots, A_n)$ such that $A_k = A$ for some $k \in \{0, 1, \ldots, n\}$. Clearly, $\boldsymbol{u}$ is a monotone game. The function $\boldsymbol{w} : \mathscr{T} \longrightarrow \mathbb{R}$ defined by

$$\boldsymbol{w}(A) = \boldsymbol{u}(A) - \boldsymbol{v}(A) \tag{15.5}$$

is monotone too. Indeed, if A and B are fuzzy coalitions with $A \leq B$, then for each chain $\mathscr{B} = (A_0, A_1, \ldots, A_n)$ with $A_k = A$ for some $k \in \{0, 1, \ldots, n\}$ the family $\mathscr{B}' = (A_0, A_1, \ldots, A_k, A_{k+1}, A_{k+2})$ with $A_{k+1} = B$ and $A_{k+2} = X$ is also a chain of fuzzy coalitions, and we have

$$\boldsymbol{u}(B) \geq \sum_{i=1}^{k} \max[\boldsymbol{v}(A_i) - \boldsymbol{v}(A_{i-1}), 0] + \max[\boldsymbol{v}(B) - \boldsymbol{v}(A), 0]$$

i.e., $\boldsymbol{u}(B) \geq \boldsymbol{u}(A) + \boldsymbol{v}(B) - \boldsymbol{v}(A)$. Hence,

$$\boldsymbol{w}(B) = \boldsymbol{u}(B) - \boldsymbol{v}(B) \geq \boldsymbol{u}(A) + \boldsymbol{v}(B) - \boldsymbol{v}(A) - \boldsymbol{v}(B) \geq \boldsymbol{w}(A).$$

Clearly, $\boldsymbol{v} = \boldsymbol{u} - \boldsymbol{w}$. Conversely, if $\boldsymbol{v} = \boldsymbol{u} - \boldsymbol{w}$, where $\boldsymbol{u}$ and $\boldsymbol{w}$ are monotone games, then

$$\|\boldsymbol{v}\| \leq \|\boldsymbol{u}\| + \|\boldsymbol{w}\| = \boldsymbol{u}(X) + \boldsymbol{w}(X) < \infty,$$

showing that $\boldsymbol{v}$ is of bounded variation. □

15.6 Definition. Let $\boldsymbol{v}$ be a game of bounded variation. The monotone game $\boldsymbol{u}$ defined by (15.4) is called *upper variation of* $\boldsymbol{v}$. The monotone game $\boldsymbol{w}$ defined by (15.5) is called *lower variation of* $\boldsymbol{v}$. The game

$$\boldsymbol{v}^* = \boldsymbol{u} + \boldsymbol{w}$$

is called *variation of* $\boldsymbol{v}$.

15.7 If Q is a linear subspace of $GAMES_{\mathscr{T}}$, then Q^+ will denote the class of monotone games contained in Q. Lemma 15.5 shows that

$$FBV_{\mathscr{T}} = FBV^+_{\mathscr{T}} - FBV^+_{\mathscr{T}}.$$

Clearly,

$$GAMES^+_{\mathscr{T}} = FBV^+_{\mathscr{T}}.$$

Lemma. *If $\boldsymbol{v}$ is a game of bounded variation then*

$$\|\boldsymbol{v}\| = \inf\{\boldsymbol{u}(X) + \boldsymbol{w}(X) \mid \boldsymbol{u}, \boldsymbol{w} \in FBV^+_{\mathscr{T}} \text{ and } \boldsymbol{v} = \boldsymbol{u} - \boldsymbol{w}\}. \tag{15.6}$$

PROOF: Let $\boldsymbol{v}$ be in $FBV_{\mathscr{T}}$ and let α be the value of the infimum in (15.6). For each chain of fuzzy coalitions $\mathscr{B} = (A_0, A_1, \ldots, A_n)$ we have:

$$\|\boldsymbol{v}\|_{\mathscr{B}} \leq \|\boldsymbol{u} - \boldsymbol{w}\|_{\mathscr{B}} + \|\boldsymbol{w}\|_{\mathscr{B}} = \boldsymbol{u}(X) + \boldsymbol{w}(X)$$

whenever $\boldsymbol{u}, \boldsymbol{w} \in FBV^{+}_{\mathscr{T}}$ and $\boldsymbol{v} = \boldsymbol{u} - \boldsymbol{w}$. Hence, $\|\boldsymbol{v}\| \leq \alpha$ because of Lemma 15.5. Now let $\boldsymbol{u}$ and $\boldsymbol{w}$ be the monotone games defined by (15.4) and (15.5), respectively. For each $\varepsilon > 0$ there exists a chain $\mathscr{B}(\varepsilon) = (A_0, A_1, \ldots, A_n)$ such that

$$\sum_{i=1}^{n} \max[\boldsymbol{v}(A_i) - \boldsymbol{v}(A_{i-1}), 0] \geq \boldsymbol{u}(X) - \varepsilon.$$

Let $\mathscr{P}$ be the subchain of $\mathscr{B}(\varepsilon)$ consisting of all links (A_{i-1}, A_i) with $\boldsymbol{v}(A_i) - \boldsymbol{v}(A_{i-1}) \geq 0$, and let $\mathscr{N}$ be the subchain of the remaining links of $\mathscr{B}(\varepsilon)$. Then we have

$$\|\boldsymbol{v}\|_{\mathscr{B}(\varepsilon)} \geq \sum_{(A_{i-1}, A_i) \in \mathscr{P}} \max[\boldsymbol{v}(A_i) - \boldsymbol{v}(A_{i-1}), 0] \geq \boldsymbol{u}(X) - \varepsilon.$$

From

$$\|\boldsymbol{v}\|_{\mathscr{B}(\varepsilon)} = \|\boldsymbol{v}\|_{\mathscr{P}} + \|\boldsymbol{v}\|_{\mathscr{N}}$$

and

$$\|\boldsymbol{v}\|_{\mathscr{P}} - \|\boldsymbol{v}\|_{\mathscr{N}} = \sum_{i=1}^{n} (\boldsymbol{v}(A_i) - \boldsymbol{v}(A_{i-1})) = \boldsymbol{v}(X)$$

it follows that

$$\|\boldsymbol{v}\| \geq \|\boldsymbol{v}\|_{\mathscr{B}(\varepsilon)} \geq 2 \cdot \boldsymbol{u}(X) - 2\varepsilon - \boldsymbol{v}(X) = \boldsymbol{u}(X) + \boldsymbol{w}(X) - 2c.$$

Therefore, we have $\|\boldsymbol{v}\| \geq \alpha - 2\varepsilon$, and taking the limit $\varepsilon \to 0$ we get $\|\boldsymbol{v}\| \geq \alpha$. □

15.8 An analysis of the proof of Lemma 15.7 yields the following:

Corollary. *If $\boldsymbol{v}$ is a game of bounded variation then there exist monotone games $\boldsymbol{u}$, $\boldsymbol{w}$, namely, the upper and lower variation of $\boldsymbol{v}$, respectively, such that*

$$\begin{aligned} \boldsymbol{v} &= \boldsymbol{u} - \boldsymbol{w}, \\ \|\boldsymbol{v}\| &= \boldsymbol{u}(X) + \boldsymbol{w}(X). \end{aligned}$$

15.9 The topological structure of the normed space $FBV_{\mathscr{T}}$ is specified in the following way:

Proposition. *$FBV_{\mathscr{F}}$ is a Banach algebra with the total variation as its norm.*

PROOF: First, we show that the normed linear space $FBV_{\mathscr{F}}$ is complete. To this end consider a Cauchy sequence $(\boldsymbol{v}_n)_{n\in\mathbb{N}}$ in $FBV_{\mathscr{F}}$ (with respect to the total variation). Since for all $m, n \in \mathbb{N}$ and each $A \in \mathscr{F}$ we have

$$|\boldsymbol{v}_n(A) - \boldsymbol{v}_m(A)| \leq \|\boldsymbol{v}_n - \boldsymbol{v}_m\|$$

it follows that $(\boldsymbol{v}_n)_{n\in\mathbb{N}}$ is a Cauchy sequence in $\mathbb{R}$, hence converging to a real number $\boldsymbol{v}(A)$.

The game $\boldsymbol{v}$ defined by $A \to \boldsymbol{v}(A)$ is of bounded variation. Indeed, there exists an $n_1 \in \mathbb{N}$ such that $\|\boldsymbol{v}_n - \boldsymbol{v}_{n_1}\| \leq 1$ whenever $n \geq n_1$. Thus for each chain $\mathscr{B}$ of fuzzy coalitions we have

$$\|\boldsymbol{v}_n\|_{\mathscr{B}} - \|\boldsymbol{v}_{n_1}\| \leq \|\boldsymbol{v}_n\|_{\mathscr{B}} - \|\boldsymbol{v}_{n_1}\|_{\mathscr{B}} \leq \|\boldsymbol{v}_n - \boldsymbol{v}_{n_1}\|_{\mathscr{B}} \leq 1$$

whenever $n \geq n_1$. Hence, $\|\boldsymbol{v}_n\|_{\mathscr{B}} \leq 1 + \|\boldsymbol{v}_{n_1}\|$ for all $n \geq n_1$. Taking the limit for $n \to \infty$, it follows that $\|\boldsymbol{v}\|_{\mathscr{B}} \leq 1 + \|\boldsymbol{v}_{n_1}\|$ for every chain of fuzzy coalitions $\mathscr{B}$. Therefore, $\|\boldsymbol{v}\| \leq 1 + \|\boldsymbol{v}_{n_1}\| < \infty$, showing that the game $\boldsymbol{v}$ belongs to $FBV_{\mathscr{F}}$. Now it can be easily seen that $(\boldsymbol{v}_n)_{n\in\mathbb{N}}$ converges to $\boldsymbol{v}$ in $FBV_{\mathscr{F}}$, and this implies that $FBV_{\mathscr{F}}$ is complete. In order to show that $FBV_{\mathscr{F}}$ is an algebra, let $\boldsymbol{v}'$, $\boldsymbol{v}''$ be games in $FBV_{\mathscr{F}}$ and let $\boldsymbol{u}'$, $\boldsymbol{u}''$ and $\boldsymbol{w}'$, $\boldsymbol{w}''$ be their upper and lower variations, respectively. Then,

$$\begin{aligned}
\|\boldsymbol{v}' \cdot \boldsymbol{v}''\| &= \|(\boldsymbol{u}' - \boldsymbol{w}') \cdot (\boldsymbol{u}'' - \boldsymbol{w}'')\| \\
&\leq \boldsymbol{u}'(X) \cdot \boldsymbol{u}''(X) + \boldsymbol{w}'(X) \cdot \boldsymbol{w}''(X) + \boldsymbol{u}'(X) \cdot \boldsymbol{w}''(X) + \boldsymbol{w}'(X) \cdot \boldsymbol{u}''(X) \\
&\leq \|\boldsymbol{v}'\| \cdot \|\boldsymbol{v}''\|,
\end{aligned}$$

thus completing the proof. □

15.10 In combination with Theorem 10.2, Proposition 9.4(ii) shows that finite T_∞-measures on $\mathscr{F}$ are necessarily elements of $FBV_{\mathscr{F}}$. According to Proposition 15.9, all positive integer powers of T_∞-measures on $\mathscr{F}$ also belong to $FBV_{\mathscr{F}}$. In general, finitely T_∞-additive functions $\boldsymbol{m}$ on $\mathscr{F}$ can be written as a difference of monotone T_∞-additive functions $\boldsymbol{m}^+$ and $\boldsymbol{m}^-$ (see Proposition 9.4(i)), but the finitely T_∞-additive functions $\boldsymbol{m}^+$ and $\boldsymbol{m}^-$ are not necessarily finite even if $\boldsymbol{m}$ is finite. Therefore, a finitely T_∞-additive function does not necessarily belong to $FBV_{\mathscr{F}}$ (even if the function assumes finite values only). The set of finitely T_∞-additive functions which belong to $FBV_{\mathscr{F}}$ is denoted by $FBA_{\mathscr{F}}$ or, simply, FBA if there is no possibility of confusion. $FBA_{\mathscr{F}}$ is a closed (and, therefore, complete) Banach subspace of $FBV_{\mathscr{F}}$. The total variation norm of a function $\boldsymbol{m} \in FBA_{\mathscr{F}}$ can be computed using the variation of the finitely T_∞-additive function $\boldsymbol{m}^*$. We can state this more precisely as follows:

Proposition. *If $\boldsymbol{m} \in FBA_{\mathcal{T}}$ then for each $A \in \mathcal{T}$*

$$\boldsymbol{m}^*(A) = \sup\left\{\sum_{i=1}^{k} |\boldsymbol{m}(A_i)| \mid (A_1, A_2, \ldots, A_k) \in \mathcal{T}_\infty(A)\right\} \tag{15.7}$$

where $\mathcal{T}_\infty(A)$ denotes the family of all finite T_∞-partitions of A consisting of elements of $\mathcal{T}$. Moreover we have

$$\|\boldsymbol{m}\| = \boldsymbol{m}^*(X). \tag{15.8}$$

PROOF: Let A be a fuzzy coalition and $\boldsymbol{m}'(A)$ the right-hand side of (15.7). Denote

$$\begin{aligned} \boldsymbol{m}_+(A) &= \frac{1}{2} \cdot (\boldsymbol{m}(A) + \boldsymbol{m}'(A)), \\ \boldsymbol{m}_-(A) &= \frac{1}{2} \cdot (\boldsymbol{m}'(A) - \boldsymbol{m}(A)). \end{aligned}$$

To show that (15.7) holds, it suffices to prove that $\boldsymbol{m}^+(A) = \boldsymbol{m}_+(A)$ and $\boldsymbol{m}^-(A) = \boldsymbol{m}_-(A)$. To this end let $M \in \mathcal{T}(A)$. Then

$$\begin{aligned} 2 \cdot \boldsymbol{m}(M) &= \boldsymbol{m}(M) + (\boldsymbol{m}(A) - \boldsymbol{m}(A \ominus M)) \\ &\leq \boldsymbol{m}(A) + |\boldsymbol{m}(A \ominus M)| + |\boldsymbol{m}(M)| \\ &\leq \boldsymbol{m}'(A) + \boldsymbol{m}(A). \end{aligned}$$

Hence,

$$\boldsymbol{m}(M) \leq \frac{1}{2} \cdot (\boldsymbol{m}'(A) + \boldsymbol{m}(A)) = \boldsymbol{m}_+(A)$$

for all $M \in \mathcal{T}(A)$. This implies that $\boldsymbol{m}^+(A) \leq \boldsymbol{m}_+(A)$. For $\varepsilon > 0$ there exists a family $(M_1, M_2, \ldots, M_k) \in \mathcal{T}_\infty(A)$ such that

$$\sum_{i=1}^{k} |\boldsymbol{m}(M_i)| \geq \boldsymbol{m}'(A) - \varepsilon.$$

Therefore, we have

$$2 \cdot \boldsymbol{m}_+(A) - \varepsilon = \boldsymbol{m}(A) + \boldsymbol{m}'(A) - \varepsilon \leq \boldsymbol{m}(A) + \sum_{i=1}^{k} |\boldsymbol{m}(M_i)|. \tag{15.9}$$

Without loss of generality, we may assume that there is a $p \in \{0, 1, \ldots, k\}$ such that

$$\begin{aligned} &\boldsymbol{m}(M_i) \geq 0 \quad \text{if} \quad i \in \{0, 1, \ldots, p\}, \\ &\boldsymbol{m}(M_i) < 0 \quad \text{if} \quad i \in \{p+1, p+2, \ldots, k\}. \end{aligned}$$

Then the right-hand side of the inequality in (15.9) can be rewritten as

$$\begin{aligned}
\boldsymbol{m}(A)+\sum_{i=1}^{k}|\boldsymbol{m}(M_i)| &= \sum_{i=1}^{p}\boldsymbol{m}(M_i)-\sum_{i=p+1}^{k}\boldsymbol{m}(M_i)+\boldsymbol{m}(A)\\
&= \boldsymbol{m}(\mathbf{S}_{\infty}{}_{i=1}^{p} M_i)-\boldsymbol{m}(\mathbf{S}_{\infty}{}_{i=p+1}^{k} M_i)+\boldsymbol{m}(\mathbf{S}_{\infty}{}_{i=1}^{k} M_i)\\
&= 2\cdot m(\mathbf{S}_{\infty}{}_{i=1}^{p} M_i)\\
&\leq 2\cdot \boldsymbol{m}^{+}(A)
\end{aligned}$$

because of

$$\mathbf{S}_{\infty}{}_{i=1}^{p} M_i \in \mathscr{T}(A).$$

As a consequence,

$$2\cdot \boldsymbol{m}_{+}(A)-\varepsilon \leq 2\cdot \boldsymbol{m}^{+}(A).$$

Taking the limit $\varepsilon \to 0$ we get $\boldsymbol{m}_{+}(A) \leq \boldsymbol{m}^{+}(A)$. Hence, $\boldsymbol{m}_{+}(A)=\boldsymbol{m}^{+}(A)$. Since

$$\boldsymbol{m}^{-}(A)=\boldsymbol{m}^{+}(A)-\boldsymbol{m}(A)=\boldsymbol{m}_{+}(A)-\boldsymbol{m}(A)=\boldsymbol{m}_{-}(A),$$

(15.7) is proved.

Now consider a chain $\mathscr{B}=(A_0,A_1,\ldots,A_n)$ of fuzzy coalitions. Then we have

$$\|\boldsymbol{m}\|_{\mathscr{B}}=\sum_{i=1}^{n}|\boldsymbol{m}(A_i\ominus A_{i-1})|,$$

and $(A_1\ominus A_0, A_2\ominus A_1,\ldots,A_n\ominus A_{n-1})\in \mathscr{T}_{\infty}(X)$. Thus we have

$$\|\boldsymbol{m}\|_{\mathscr{B}} \leq \boldsymbol{m}'(X)=\boldsymbol{m}^{*}(X),$$

i.e., $\|\boldsymbol{m}\| \leq \boldsymbol{m}^{*}(X)$. If $(A_1,A_2,\ldots,A_n)\in \mathscr{T}_{\infty}(A)$, then the family of fuzzy coalitions $(A_1',A_2',\ldots,A_n')$ defined by

$$A_j'=\mathbf{S}_{\infty}{}_{i=1}^{j} A_i$$

forms a chain $\mathscr{B}$ in $\mathscr{T}$ and we have

$$\sum_{i=1}^{n}|\boldsymbol{m}(A_i)|=\sum_{i=1}^{n}|\boldsymbol{m}(A_i')-\boldsymbol{m}(A_{i-1}')|=\|\boldsymbol{m}\|_{\mathscr{B}}\leq\|\boldsymbol{m}\|.$$

Thus $\boldsymbol{m}^{*}(X)\leq\|\boldsymbol{m}\|$, and by virtue of (15.7) the proposition is proved. □

15.11 An interesting and important facet of the topology of $FBV_{\mathcal{T}}$ is that the variation norm is somewhat *conditioned* by the nonatomic T_∞-measures on $\mathcal{T}$ (if such T_∞-measures exist at all). To be precise, we have the following result:

Lemma. *If $\boldsymbol{v}$ is a game and $\boldsymbol{m}$ is a nonnegative, nonatomic T_∞-measure on $\mathcal{T}$, then for each $\eta > 0$*

$$\|\boldsymbol{v}\| = \sup_{\mathcal{B}}(\|\boldsymbol{v}\|_{\mathcal{B}}),$$

where the supremum is taken over all chains $\mathcal{B} = (A_0, A_1, \ldots, A_n)$ such that

$$\boldsymbol{m}(A_i \ominus A_{i-1}) \leq \eta$$

whenever (A_{i-1}, A_i) is a link in $\mathcal{B}$.

PROOF: If $\mathcal{B}$ is a chain of fuzzy coalitions and if it contains a link (A_{i-1}, A_i) such that $\boldsymbol{m}(A_i \ominus A_{i-1}) > \eta$, then $\mathcal{B}$ can be embedded into a chain $\mathcal{B}'$ such that $\boldsymbol{m}(B \ominus A) \leq \eta$ for all links (A, B) in $\mathcal{B}'$. This can be verified in the following way: Choose $W_1 \in \mathcal{T}(A_i \ominus A_{i-1})$ such that

$$\boldsymbol{m}(W_1) = \frac{1}{2} \cdot \boldsymbol{m}(A_i \ominus A_{i-1})$$

and include the fuzzy set $W_1\, \mathbf{S}_\infty\, A_{i-1}$ into $\mathcal{B}$. We repeat this procedure for each link (A_{i-1}, A_i) in $\mathcal{B}$ with $\boldsymbol{m}(A_{i-1}, A_i) > \eta$. This is possible because of the monotonicity of $\boldsymbol{m}$. If the chain obtained in this manner does not satisfy the assumption of the lemma, then repeat the whole procedure replacing $\mathcal{B}$ by the newly obtained chain. After finitely many steps of this kind, we obtain the required chain $\mathcal{B}'$. Clearly $\|\boldsymbol{v}\|_{\mathcal{B}} \leq \|\boldsymbol{v}\|_{\mathcal{B}'}$, which means that in the definition of $\|\boldsymbol{v}\|$ it is sufficient to consider only those chains which satisfy the condition of the lemma. □

16 The Space *FAC*

16.1 In this section we study a closed subspace of $FBV_{\mathcal{T}}$ which we shall call $FAC_{\mathcal{T}}$. The space $FAC_{\mathcal{T}}$ will be crucially involved when showing the existence of the *Aumann–Shapley value* for a class of games with fuzzy coalitions. Before giving a precise definition of the space $FAC_{\mathcal{T}}$, we need the following preparatory result:

Lemma. *If $\boldsymbol{m}$ and $\boldsymbol{v}$ are T_∞-measures on the T_∞-tribe $\mathcal{T}$ and if $\boldsymbol{m}$ is monotone, then $\boldsymbol{v}$ is $\boldsymbol{m}$-absolutely continuous if and only if the following condition (*) is satisfied:*

(*) *for each $\varepsilon > 0$ there exists a $\delta(\varepsilon) > 0$ such that for each chain $\mathcal{B}$ of fuzzy coalitions and for each subchain $\mathcal{C}$ of $\mathcal{B}$*

$$\|\boldsymbol{v}\|_{\mathcal{C}} < \varepsilon \quad \textit{whenever} \quad \|\boldsymbol{m}\|_{\mathcal{C}} < \delta(\varepsilon). \tag{16.1}$$

PROOF: If (*) holds and if E is a fuzzy coalition with $\boldsymbol{m}(E) = 0$, then the subchain $\mathscr{C} = \{(\emptyset, E)\}$ of the chain $(\emptyset, E, X)$ satisfies $\|\boldsymbol{m}\|_{\mathscr{C}} < \delta(\frac{1}{k})$ for each $k \in \mathbb{N}$, and we get $\|\boldsymbol{v}\|_{\mathscr{C}} < \frac{1}{k}$ which implies $\boldsymbol{v}(E) = 0$. Hence, $\boldsymbol{v}$ is $\boldsymbol{m}$-absolutely continuous.

Conversely, let $\boldsymbol{v}$ be an $\boldsymbol{m}$-absolutely continuous T_∞-measure on $\mathscr{T}$ and assume that (*) does not hold. Then there is an $\varepsilon_0 > 0$ such that for each $k \in \mathbb{N}$ there exists a chain $\mathscr{B}_k = (A_0^k, A_1^k, \ldots, A_{h(k)}^k)$ containing a subchain $\mathscr{C}_k$ such that $\|\boldsymbol{v}\|_{\mathscr{C}_k} < \frac{1}{k}$ and $\|\boldsymbol{v}\|_{\mathscr{C}_k} \geq \varepsilon_0$. Putting

$$E = \liminf_{k\to\infty} E_k,$$

where

$$E_k = \mathbf{S}_{\infty\,(A_{i-1}^k, A_i^k)\in\mathscr{C}_k} (A_i^k \ominus A_{i-1}^k),$$

we derive

$$\boldsymbol{m}(E_k) \leq \sum_{(A_{i-1}^k, A_i^k)\in\mathscr{C}_k} \boldsymbol{m}(A_i^k \ominus A_{i-1}^k) = \|\boldsymbol{m}\|_{\mathscr{C}_k} \leq \frac{1}{k}.$$

Since $\boldsymbol{m}$ is nonnegative and finite, it follows that

$$0 \leq \boldsymbol{m}(E) = \boldsymbol{m}(\liminf_{k\to\infty} E_k) = \liminf_{k\to\infty} \boldsymbol{m}(E_k) = 0,$$

i.e., $\boldsymbol{m}(E) = 0$, and, because of $\boldsymbol{v} \ll \boldsymbol{m}$, $\boldsymbol{v}^*(E) = 0$. On the other hand we have

$$\boldsymbol{v}^*(E) \geq \sum_{(A_{i-1}^k, A_i^k)\in\mathscr{C}_k} |\boldsymbol{v}(A_i^k \ominus A_{i-1}^k)| = \|\boldsymbol{v}\|_{\mathscr{C}_k} \geq \varepsilon_0 > 0$$

by virtue of Proposition 15.10, and this contradicts our assumption. □

16.2 Condition (*) allows the concept of absolute continuity to be extended from T_∞-measures to games.

Definition.

(i) Let $\boldsymbol{m}$ be a nonnegative T_∞-measure and $\boldsymbol{v}$ a game on $\mathscr{T}$. If condition (*) holds, then $\boldsymbol{v}$ is said to be $\boldsymbol{m}$-*absolutely continuous* (briefly $\boldsymbol{v} \ll \boldsymbol{m}$).

(ii) The game $\boldsymbol{v}$ is called NA-*absolutely continuous* if there exists a nonnegative finite nonatomic T_∞-measure $\boldsymbol{m}$ such that $\boldsymbol{v} \ll \boldsymbol{m}$. The family of all NA-absolutely continuous games is denoted $FAC_{\mathscr{T}}$ or simply FAC when there is no possibility of confusion.

16.3 The following is an extension of a result in [AUMANN & SHAPLEY 1974, Proposition 5.2] to arbitrary T_∞-tribes.

Proposition. *$FAC_{\mathcal{T}}$ is a closed linear subspace of $FBV_{\mathcal{T}}$.*

PROOF: First, we show that $FAC_{\mathcal{T}}$ is a linear subspace of $FBV_{\mathcal{T}}$. It is obvious that, whenever $\boldsymbol{v}_1$ and $\boldsymbol{v}_2$ are games in $FAC_{\mathcal{T}}$, $\boldsymbol{v}_1 - \boldsymbol{v}_2$ also belongs to $FAC_{\mathcal{T}}$. Therefore, it is sufficient to show that $FAC_{\mathcal{T}} \subseteq FBV_{\mathcal{T}}$. To this end, let $\boldsymbol{v} \in FAC_{\mathcal{T}}$ and $\boldsymbol{m}$ be a monotone T_∞-measure such that $\boldsymbol{v} \ll \boldsymbol{m}$. According to condition (*), there exists a $\delta > 0$ such that for each chain $\mathscr{B}$ of fuzzy coalitions and for each subchain $\mathscr{C}$ of $\mathscr{B}$ we have $\|\boldsymbol{v}\|_{\mathscr{C}} < 1$ whenever $\|\boldsymbol{m}\|_{\mathscr{C}} < \delta$. If $\mathscr{B}$ is a chain of fuzzy coalitions such that for each link (A_{i-1}, A_i) of $\mathscr{B}$ we have $\boldsymbol{m}(A_i \ominus A_{i-1}) < \delta$, then let $k_0 \in \mathbb{N}$ be the smallest number such that

$$\sum_{i=0}^{k_0} \boldsymbol{m}(A_i \ominus A_{i-1}) \geq \frac{\delta}{2},$$

and let $\mathscr{C}_0$ be the subchain of $\mathscr{B}$ which consists of all links (A_{i-1}, A_i) with $i \in \{1, 2, \ldots, k_0\}$. Let $k_1 \in \{k_0 + 1, k_0 + 2, \ldots\}$ be the smallest number such that

$$\sum_{i=k_0+1}^{k_1} \boldsymbol{m}(A_i \ominus A_{i-1}) \geq \frac{\delta}{2},$$

Denote by $\mathscr{C}_1$ the subchain of $\mathscr{B}$ consisting of all links (A_{i-1}, A_i) with $i \in \{k_0 + 1, k_0 + 2, \ldots, k_1\}$. Continuing in this way, subchains $\mathscr{C}_0, \mathscr{C}_1, \ldots, \mathscr{C}_t$ of $\mathscr{B}$ are obtained where $\mathscr{C}_j$ consists of all links (A_{i-1}, A_i) of $\mathscr{B}$ with $i \in \{k_{j-1} + 1, k_{j-1} + 2, \ldots, k_j\}$ and $k_j \in \{k_{j-1} + 1, k_{j-1} + 2, \ldots\}$ is the smallest number such that

$$\sum_{i=k_{j-1}+1}^{k_j} \boldsymbol{m}(A_i \ominus A_{i-1}) \geq \frac{\delta}{2},$$

Let $\mathscr{C}_{t+1}$ be the subchain of $\mathscr{B}$ consisting of all links of $\mathscr{B}$ which are not contained in any one of the subchains $\mathscr{C}_1, \mathscr{C}_2, \ldots, \mathscr{C}_t$. If $s \leq t$, then

$$\|\boldsymbol{m}\|_{\mathscr{C}_{t+1}} < \frac{\delta}{2} \leq \|\boldsymbol{m}\|_{\mathscr{C}_s} \leq \|\boldsymbol{m}\|_{\mathscr{B}} \leq \delta.$$

Consider the fuzzy coalition defined by

$$H_s = \mathop{\mathbf{S}_\infty}_{i=k_{s-1}+1}^{k_s} (A_i \ominus A_{i-1}).$$

Then we have $\boldsymbol{m}(H_s) = \|\boldsymbol{m}\|_{\mathscr{C}_s}$ for all $s \in \{1, 2, \ldots, t+1\}$. Since the family of fuzzy sets $(H_s)_{s \in \{2,3,\ldots,t\}}$ forms a finite T_∞-partition of X, it follows that

$$\boldsymbol{m}(X) = \boldsymbol{m}(H_{t+1}) + \sum_{s=1}^{t} \boldsymbol{m}(H_s) \geq t \cdot \frac{\delta}{2}$$

and, subsequently,

$$t \leq \frac{2 \cdot \boldsymbol{m}(\mathrm{X})}{\delta}.$$

Since $\|\boldsymbol{m}\|_{\mathscr{C}_s} < \delta$, we obtain $\|\boldsymbol{v}\|_{\mathscr{C}_s} \leq 1$ whenever $s \in \{1, 2, \ldots, t+1\}$ and, therefore,

$$\|\boldsymbol{v}\|_{\mathscr{B}} = \sum_{s=1}^{t+1} \leq \|\boldsymbol{v}\|_{\mathscr{C}_s} \leq (t+1) \leq \frac{2 \cdot \boldsymbol{m}(\mathrm{X})}{\delta} + 1 < \infty.$$

If $\boldsymbol{m}$ is a nonatomic T_∞-measure, then, exactly as in the proof of Lemma 15.11, each chain of fuzzy coalitions $\mathscr{C}$ can be embedded into a chain of fuzzy coalitions $\mathscr{B}$ such that we have $\boldsymbol{m}(A_i \ominus A_{i-1}) < \delta$ whenever (A_{i-1}, A_i) is a link of $\mathscr{B}$. Using Lemma 15.11 it follows that $\|\boldsymbol{v}\|$ is finite, i.e., $\boldsymbol{v} \in FBV_{\mathscr{T}}$. Now we prove that $FAC_{\mathscr{T}}$ is a closed subspace of $FBV_{\mathscr{T}}$. Let $(\boldsymbol{v}_n)_{n \in \mathbb{N}}$ be a sequence of games in $FAC_{\mathscr{T}}$ and assume that it converges to $\boldsymbol{v}$ (with respect to the total variation). Then $\boldsymbol{v} \in FBV_{\mathscr{T}}$ since each $\boldsymbol{v}_n$ is contained in $FBV_{\mathscr{T}}$ and because of the closedness of $FBV_{\mathscr{T}}$. Then for each $n \in \mathbb{N}$ there exists a nonatomic nonnegative finite T_∞-measure $\boldsymbol{m}_{n'}$ such that $\boldsymbol{v}_n \ll \boldsymbol{m}_n$. Consider the function $\boldsymbol{m} : \mathscr{T} \longrightarrow \mathbb{R}$ defined by

$$\boldsymbol{m}(A) = \sum_{n=1}^{\infty} \frac{\boldsymbol{m}_n(A)}{2^n \cdot \boldsymbol{m}_n(X)}.$$

A standard argument shows that $\boldsymbol{m}$ is a nonatomic nonnegative finite T_∞-measure. It remains to show that $\boldsymbol{v} \ll \boldsymbol{m}$. Choose $\varepsilon > 0$. For each $n \in \mathbb{N}$ there exists a $\delta_n(\varepsilon) > 0$ such that $\|\boldsymbol{m}_n\|_{\mathscr{C}} < \delta_n(\varepsilon)$ implies $\|\boldsymbol{v}_n\|_{\mathscr{C}} < \varepsilon$ whenever $\mathscr{C}$ is a subchain of a chain of fuzzy coalitions. Also, a number $n(\varepsilon) \in \mathbb{N}$ exists such that $\|\boldsymbol{v}_n - \boldsymbol{v}\| < \frac{\varepsilon}{2}$ whenever $n \geq n(\varepsilon)$. Denote

$$\delta(\varepsilon) = \frac{\delta_{n(\varepsilon)}(\varepsilon)}{2^{n(\varepsilon)} \cdot \boldsymbol{m}_{n(\varepsilon)}(X)}.$$

If $\mathscr{B}$ is a chain of fuzzy coalitions and $\mathscr{C}$ is a subchain of it such that $\|\boldsymbol{m}\|_{\mathscr{C}} < \delta(\varepsilon)$, then for each $n \in \mathbb{N}$ we have

$$\begin{aligned}
\delta(\varepsilon) \quad &> \quad \sum_{(A_{i-1}, A_i) \in \mathscr{C}_k} \boldsymbol{m}(A_i \ominus A_{i-1}) \\
&= \quad \sum_{(A_{i-1}, A_i) \in \mathscr{C}_k} \sum_{k=1}^{\infty} \frac{\boldsymbol{m}_k(A_i \ominus A_{i-1})}{2^k \cdot \boldsymbol{m}_k(X)} \\
&\geq \quad \frac{\|\boldsymbol{m}\|_{\mathscr{C}}}{2^{n(\varepsilon)} \cdot \boldsymbol{m}_{n(\varepsilon)}(X)}.
\end{aligned}$$

Hence,

$$\|\boldsymbol{v}_{n(\varepsilon)}\|_{\mathscr{C}} < \frac{\varepsilon}{2 \cdot \boldsymbol{m}_{n(\varepsilon)}(X)},$$

and this implies

$$\|v\|_{\mathscr{C}} \leq \|v_{n(\varepsilon)} - v\|_{\mathscr{C}} + \|v_{n(\varepsilon)}\|_{\mathscr{C}} \leq \|v_{n(\varepsilon)} - v\| + \|v_{n(\varepsilon)}\|_{\mathscr{C}} < \varepsilon.$$

Therefore, $\|v\|_{\mathscr{C}} < \varepsilon$ whenever $\|m\|_{\mathscr{C}} < \delta(\varepsilon)$, thus completing the proof. □

17 The Space *pFNA*

17.1 We shall denote the family of finite nonatomic T_∞-measures on a T_∞-tribe $\mathscr{T}$ by $FNA_{\mathscr{T}}$, or simply FNA if there is no possibility of confusion. Obviously, $FNA_{\mathscr{T}}$ is a linear subspace of $FBV_{\mathscr{T}}$. Denote $pFNA_{\mathscr{T}}$ the closed subspace of $GAMES_{\mathscr{T}}$ spanned by all positive integer powers of T_∞-measures in $FNA^+_{\mathscr{T}}$, i.e., the closure of the linear hull of $\{m^k \mid m \in FNA^+_{\mathscr{T}}, k \in \mathbb{N}\}$.

Proposition. *The space $pFNA_{\mathscr{T}}$ is a closed linear subspace of $FAC_{\mathscr{T}}$.*

PROOF: It suffices to show that for each m in $FNA^+_{\mathscr{T}}$ and for each $k \in \mathbb{N}$ we have $m^k \in FAC_{\mathscr{T}}$. Let $\mathscr{B} = (A_0, A_1, \ldots, A_n)$ be a chain of fuzzy coalitions and $\mathscr{C}$ a subchain of $\mathscr{B}$. Then

$$\begin{aligned} \|m^k\|_{\mathscr{C}} &= \sum_{A_{i-1}, A_i \in \mathscr{C}} |m^k(A_i) - m^k(A_{i-1})| \\ &= \sum_{A_{i-1}, A_i \in \mathscr{C}} m(A_i \ominus A_{i-1}) \cdot \sum_{j=0}^{k-1} m^{k-j-1}(A_i) \cdot m^j(A_{i-1}) \\ &\leq k \cdot m^{k-1}(X) \cdot \|m\|_{\mathscr{C}}. \end{aligned}$$

Therefore,

$$\|m^k\|_{\mathscr{C}} < \varepsilon \quad \text{whenever} \quad \|m\|_{\mathscr{C}} < \frac{\varepsilon}{k \cdot m^{k-1}(X)},$$

which means $m^k \ll m$ and, consequently, $m^k \in FAC_{\mathscr{T}}$. This, together with Proposition 16.3, completes the proof. □

17.2 In the case where $\mathscr{T}$ is a σ-algebra of crisp subsets of X the space $pFNA_{\mathscr{T}}$ was studied by AUMANN& SHAPLEY [1974]. The following result is, in principle, their Theorem C restated for arbitrary T_∞-tribes instead of σ-algebras of crisp subsets of X. We produce a full proof of this result since our more general context requires some technical changes because of dealing with fuzzy sets and, most important, some steps are based on Theorem 14.2 rather than the classical Liapounoff theorem involved in the original proof by AUMANN & SHAPLEY [1974, Section 5]. Recall that Theorem

14.2 guarantees that $\mathcal{R}(\boldsymbol{m})$ is an interval whenever $\boldsymbol{m} \in FNA_{\mathcal{T}}$. Therefore, the following statement is meaningful.

Theorem. *Let $\boldsymbol{m}$ be a monotone nonatomic finite T_∞-measure on $\mathcal{T}$, $f : \mathcal{R}(\boldsymbol{m}) \longrightarrow \mathbb{R}$ a function such that $f(0) = 0$, and denote $\boldsymbol{v} = f \circ \boldsymbol{m}$. The game $\boldsymbol{v}$ belongs to $pFNA_{\mathcal{T}}$ if and only if the function f is absolutely continuous.*

PROOF: First, assume that f is absolutely continuous on the interval $\mathcal{R}(\boldsymbol{m})$ with $f(0) = 0$. Then $f = f_1 - f_2$, where $f_1, f_2 : \mathcal{R}(\boldsymbol{m}) \longrightarrow \mathbb{R}$ are increasing and absolutely continuous functions satisfying $f_1(0) = f_2(0) = 0$. Clearly, f_1 and f_2 are distribution functions and they uniquely define two σ-additive measures $\boldsymbol{n}_1$ and $\boldsymbol{n}_2$ on the σ-algebra of Borel subsets of $\mathcal{R}(\boldsymbol{m})$ such that $\boldsymbol{n}_i([a,b]) = f_i(b) - f_i(a)$ whenever $[a,b] \subseteq \mathcal{R}(\boldsymbol{m})$ and $i \in \{1,2\}$ (cf. [PARTHASARATHY 1978, Proposition 15.2]). Denote the Lebesgue measure on $\mathcal{R}(\boldsymbol{m})$ by λ. Then $\boldsymbol{n}_1$ and $\boldsymbol{n}_2$ are λ-absolutely continuous. The Radon–Nikodym Theorem ensures the existence of two λ-integrable functions $g_1, g_2 : \mathcal{R}(\boldsymbol{m}) \longrightarrow \mathbb{R}$ such that

$$\boldsymbol{n}_i(A) = \int_A g_i \, \mathrm{d}\lambda$$

for all Borel subsets $A \subseteq \mathcal{R}(\boldsymbol{m})$ and $i \in \{1,2\}$. Then for each $x \in \mathcal{R}(\boldsymbol{m})$ we have

$$f_i(x) = \int_{[0,x]} g_i \, \mathrm{d}\lambda.$$

Hence, denoting $g = g_1 - g_2$ we obtain

$$f(x) = f_1(x) - f_2(x). = \int_{[0,x]} g \, \mathrm{d}\lambda.$$

Since g is λ-integrable, there exists a sequence of continuous functions defined on the interval $\mathcal{R}(\boldsymbol{m})$ which converges λ-a.e. pointwise to g. Each continuous function on $\mathcal{R}(\boldsymbol{m})$ is, in turn, the uniform limit of a sequence of polynomials. Now, Cantor's diagonal procedure yields a λ-a.e. pointwise convergent sequence of polynomials $(q_n)_{n \in \mathbb{N}}$ such that $\lim_{n\to\infty} q_n = g$ λ-a.e. For each $n \in \mathbb{N}$ define $p_n : \mathcal{R}(\boldsymbol{m}) \longrightarrow \mathbb{R}$ by

$$p_n(x) = \int_{[0,x]} q_n \, \mathrm{d}\lambda.$$

Clearly, each p_n is a polynomial and

$$\lim_{n\to\infty} \int_{\mathcal{R}(\boldsymbol{m})} |g - q_n| \, \mathrm{d}\lambda = 0.$$

If $V_n(\boldsymbol{m}, f)$ denotes the total variation of $|p_n - f|$ on $\mathcal{R}(\boldsymbol{m})$, then we have

$$0 \le V_n(\boldsymbol{m}; f) \le \int_{\mathcal{R}(\boldsymbol{m})} |q_n - g| \, \mathrm{d}\lambda,$$

and thus f is the limit (in variation) of the sequence of polynomials $(p_n)_{n\in\mathbb{N}}$. For $n \in \mathbb{N}$ denote $\boldsymbol{v}_n = p_n \circ \boldsymbol{m}$. Then for each chain of fuzzy coalitions $\mathscr{B} = (A_0, A_1, \ldots, A_k)$, we have

$$\|\boldsymbol{v}_n - \boldsymbol{v}\|_{\mathscr{B}} = \sum_{i=1}^{k} |\boldsymbol{v}_n(A_i) - \boldsymbol{v}(A_i) - \boldsymbol{v}_n(A_{i-1}) + \boldsymbol{v}(A_{i-1})| \leq V_n(\boldsymbol{m}, f).$$

Thus $(\boldsymbol{v}_n)_{n\in\mathbb{N}}$ converges in variation to $\boldsymbol{v}$. Since $pFNA_{\mathscr{T}}$ is closed and each $\boldsymbol{v}_n \in pFNA_{\mathscr{T}}$, it follows that $\boldsymbol{v} \in pFNA_{\mathscr{T}}$.

Conversely, assume that $\boldsymbol{v} = f \circ \boldsymbol{m} \in pFNA_{\mathscr{T}}$. Then $\boldsymbol{v} \in FAC_{\mathscr{T}}$ (cf. Proposition 17.1) and hence there is a T_∞-measure $\boldsymbol{n} \in FNA^+_{\mathscr{T}}$ such that $\boldsymbol{v} \ll \boldsymbol{n}$. The 2-vector T_∞-measure $\boldsymbol{w} = (\boldsymbol{m}, \boldsymbol{n})$ is nonatomic and has the Darboux property (cf. Theorem 13.3). As a consequence, there exists a family $(A_r)_{r\in[0,1]}$ of fuzzy coalitions such that $\boldsymbol{w}(A_r) = r \cdot \boldsymbol{w}(X)$ for all $r \in [0,1]$, and $A_r \leq A_s$ whenever $0 \leq r < s \leq 1$. For each $\varepsilon > 0$ there is a $\delta > 0$ such that

$$\|\boldsymbol{v}\|_{\mathscr{C}} < \varepsilon \quad \text{whenever} \quad \|\boldsymbol{n}\|_{\mathscr{C}} < \delta \tag{17.1}$$

for all subchains $\mathscr{C}$ of a chain of fuzzy coalitions. If

$$0 \leq a_0 \leq b_0 \leq a_1 \leq b_1 \leq \ldots \leq a_n \leq b_n \leq \boldsymbol{m}(X)$$

are such that

$$\sum_{i=0}^{n} (b_i - a_i) < \delta \cdot \frac{\boldsymbol{m}(X)}{\boldsymbol{n}(X)},$$

then each a_i can be written as $r_i \cdot \boldsymbol{m}(X)$ for some $r_i \in [0,1]$ and each b_i can be written as $s_i \cdot \boldsymbol{m}(X)$ for some $s_i \in [0,1]$, and the pairs (A_{r_i}, A_{s_i}) form a subchain $\mathscr{C}$ of a chain of fuzzy coalitions such that

$$\|\boldsymbol{n}\|_{\mathscr{C}} = \sum_{i=1}^{n} |\boldsymbol{n}(A_{s_i}) - \boldsymbol{n}(A_{r_i})| = \frac{\boldsymbol{n}(X)}{\boldsymbol{m}(X)} \cdot \sum_{i=1}^{n} |\boldsymbol{m}(A_{s_i}) - \boldsymbol{m}(A_{r_i})| < \delta. \tag{17.2}$$

Combining (17.1) and (17.2) we get

$$\sum_{i=1}^{n} |f(b_i) - f(a_i)| = \|\boldsymbol{v}\|_{\mathscr{C}} \leq \varepsilon,$$

showing that f is absolutely continuous. □

17.3 In some further developments, the following slightly modified restatement of Lemma 7.2 of AUMANN & SHAPLEY [1974, p. 41] will be useful.

Lemma. *If $\boldsymbol{m} = (\boldsymbol{m}_1, \boldsymbol{m}_2, \ldots, \boldsymbol{m}_n)$ is an n-vector T_∞-measure on $\mathcal{T}$ and if f is a polynomial in n variables such that $f(0) = 0$, then $f \circ \boldsymbol{m}$ can be represented as a linear combination of positive integer powers of finite nonnegative T_∞-measures on $\mathcal{T}$. If $\boldsymbol{m}$ is also nonatomic, then $f \circ \boldsymbol{m}$ can be represented as a linear combination of positive integer powers of T_∞-measures in $FNA^+_{\mathcal{T}}$.*

PROOF: Since f is a polynomial in n variables, it can be written as a linear combination of monomials of its variables. If $x_1, x_2, \ldots, x_k$ are real variables, then

$$\begin{aligned} k! \cdot (x_1 \cdot x_2 \cdot \ldots \cdot x_k) &= \\ &= (x_1 + x_2 + \cdots + x_k)^k - \sum_{1 \le i \le k} (x_1 + \cdots + x_k - x_i)^k \\ &+ (x_1 + x_2 + \cdots + x_k - (x_i + x_j))^k + \cdots + (-1)^{k-1} \cdot \sum_{i=1}^{k} x_i^k. \end{aligned} \tag{17.3}$$

Indeed, the right hand side is a polynomial of degree k which vanishes whenever $x_i = 0$ for some index i, and the monomial

$$x_1 \cdot x_2 \cdot \cdots \cdot x_k$$

appears in the first term on the right hand side of (17.3) with

$$\binom{k}{1} \cdot \binom{k-1}{1} \cdot \cdots \cdot \binom{1}{1} = k!$$

as its coefficient. Formula (17.3) shows that the polynomial f can be represented as a linear combination of positive integer powers of partial sums of its variables. Each T_∞-measure $\boldsymbol{m}_i$ can be written as $\boldsymbol{m}_i = \boldsymbol{m}_i^+ - \boldsymbol{m}_i^-$ (cf. Theorem 10.2), $\boldsymbol{m}_i^+$ and $\boldsymbol{m}_i^-$ being nonnegative T_∞-measures. Since the sum of finitely many T_∞-measures is still a T_∞-measure, the first statement in the lemma follows. It can be easily seen that $\boldsymbol{m}_i$ is nonatomic if and only if $\boldsymbol{m}_i^+$ and $\boldsymbol{m}_i^-$ are nonatomic. Therefore, the second statement of the lemma follows from (17.3) too. □

17.4 Now we can formulate the following straightforward generalization of Proposition 7.1 in [AUMANN & SHAPLEY 1974].

Proposition. *If $\boldsymbol{m} = (\boldsymbol{m}_1, \boldsymbol{m}_2, \ldots, \boldsymbol{m}_n)$ is a nonatomic n-vector T_∞-measure on $\mathcal{T}$ and if $f : \mathcal{R}(\boldsymbol{m}) \longrightarrow \mathbb{R}$ is continuously differentiable and $f(0) = 0$, then $\boldsymbol{v} = f \circ \boldsymbol{m} \in pFNA_{\mathcal{T}}$.*

PROOF: This result is obtained from the proof of Proposition 7.1 in [AUMANN & SHAPLEY 1974], using T_∞-measures instead of measures and applying Theorem 14.2 instead of the Liapounoff Theorem. □

18 Diagonal Aumann–Shapley Value on *pFNA*

18.1 The aim of this section is to prove the existence of an operator on $pFNA_{\mathscr{T}}$ which satisfies the *plausibility conditions* formulated axiomatically by AUMANN & SHAPLEY [1974, p.15]. Such a value will be called an *Aumann–Shapley value* or briefly A–S value. To define the concept of an Aumann–Shapley value, some notions and notations are needed.

Let $\pi : X \longrightarrow X$ be a one-to-one mapping. We say that π is a *permutation of* $(X, \mathscr{T})$ if πA and $\pi^{-1}A$ are contained in $\mathscr{T}$ whenever $A \in \mathscr{T}$, where the functions $\pi A, \pi^{-1}A : X \longrightarrow X$ are defined as follows:

$$(\pi A)(x) = A(\pi^{-1}(x)), \tag{18.1}$$

$$(\pi^{-1}A)(x) = A(\pi(x)). \tag{18.2}$$

Obviously, a permutation of $(X, \mathscr{T})$ is always an automorphism of the measure space $(X, \mathscr{T}^{\vee})$. If $\mathscr{T}$ is not generated, the converse is not generally true: if, for instance, $X = \{a, b\}$ and $\mathscr{T} = \{A \in [0,1]^X \mid A(a) \in \{0,1\}\}$, then $\pi : X \longrightarrow X$, defined by $\pi(a) = b$ and $\pi(b) = a$, is an automorphism on $(X, \mathscr{T}^{\vee}) = (X, \mathscr{P}(X))$, but not a permutation of $(X, \mathscr{T})$ since for $A \in \mathscr{T}$, given by $A(a) = 0$ and $A(b) = \frac{1}{2}$, we have $(\pi A)(a) = \frac{1}{2}$ and, therefore, $\pi A \notin \mathscr{T}$.

If π is a permutation of $(X, \mathscr{T})$, then it induces a linear mapping $\pi_* : FBV_{\mathscr{T}} \longrightarrow FBV_{\mathscr{T}}, \boldsymbol{v} \longmapsto \pi_*\boldsymbol{v}$, as follows:

$$\pi_*\boldsymbol{v}(A) = \boldsymbol{v}(\pi A). \tag{18.3}$$

A linear subspace Q of $FBV_{\mathscr{T}}$ is called *symmetric* if $\pi_*Q \subseteq Q$ for all permutations π of $(X, \mathscr{T})$.

An operator $\varphi : Q \longrightarrow FBV_{\mathscr{T}}$ is called *positive* if it preserves monotonicity, i.e., if $\varphi\boldsymbol{v}$ is monotone whenever $\boldsymbol{v}$ is monotone. It is easily seen that $pFNA_{\mathscr{T}}$ and $FAC_{\mathscr{T}}$ are symmetric subspaces of $FBV_{\mathscr{T}}$.

18.2 Definition. Let Q be a symmetric subspace of $FBV_{\mathscr{T}}$. An *Aumann–Shapley value* on Q is a positive linear operator $\varphi : Q \longrightarrow FBA_{\mathscr{T}}$, $\boldsymbol{v} \longmapsto \varphi\boldsymbol{v}$, which satisfies the following *plausibility conditions*:

(i) *Symmetry*: if π is a permutation of $(X, \mathscr{T})$, then for each $\boldsymbol{v} \in Q$

$$\varphi(\pi_*\boldsymbol{v}) = \pi_*(\varphi\boldsymbol{v}). \tag{18.4}$$

(ii) *Efficiency*: for all $\boldsymbol{v} \in Q$

$$\varphi\boldsymbol{v}(X) = \boldsymbol{v}(X). \tag{18.5}$$

If φ is an Aumann–Shapley value on Q and $\boldsymbol{v} \in Q$, then $\varphi\boldsymbol{v}$ is called a *value of* $\boldsymbol{v}$ and, for A in $\mathscr{T}$, $\varphi\boldsymbol{v}(A)$ is called a *value of the fuzzy coalition* A in the game $\boldsymbol{v}$.

18.3 The problem of the existence of an Aumann–Shapley value on a subspace Q of $FBV_{\mathscr{T}}$ has many facets. If φ is an Aumann–Shapley value on the subspace Q of $FBV_{\mathscr{T}}$, then, for each $\boldsymbol{v} \in Q$, $\varphi\boldsymbol{v}$ can be seen as a redistribution of the worth of the coalitions in a *fair manner*. This interpretation (see [AUMANN & SHAPLEY 1974, p. 15]) goes back to SHAPLEY's [1953] original paper in which the foundations of the axiomatic theory of value were presented. From another point of view, if $\boldsymbol{v}$ is a *market game*, then we may consider the value $\varphi\boldsymbol{v}$ as a candidate for the *core* of $\boldsymbol{v}$ which would eventually define a *competitive payoff density* in $\boldsymbol{v}$ (see [AUMANN & PERLES 1965] and [AUMANN & SHAPLEY 1974, Section 32]). From a strictly theoretical point of view, the existence of an Aumann–Shapley value on Q signifies that the structure of Q and of its elements is quite *regular* since, as it will be made clear later, the Aumann–Shapley value on Q behaves in some respects like a differentiability operator.

18.4 In this book, the problem of the existence of Aumann–Shapley values in various subspaces of $FBV_{\mathscr{T}}$ is approached as a problem of extending the existing Aumann–Shapley (diagonal) value on $pFNA_{\mathscr{T}}$. The fact that an Aumann-Shapley value on $pFNA_{\mathscr{T}}$ exists is shown by the following generalization of the existence part of Theorem B of AUMANN & SHAPLEY [1974]:

Theorem. *There exists an Aumann–Shapley value φ on $pFNA_{\mathscr{T}}$ which is continuous and has norm 1, such that for each nonatomic vector T_∞-measure $\boldsymbol{m}$ on $\mathscr{T}$ and for each continuously differentiable function $f : \mathcal{R}(\boldsymbol{m}) \longrightarrow \mathbb{R}$ with $f(0) = 0$ the following diagonal formula holds:*

$$\varphi(f \circ \boldsymbol{m})(A) = \int_0^1 f_{\boldsymbol{m}(A)}(t \cdot \boldsymbol{m}(X))\, dt, \tag{18.6}$$

where $f_{\boldsymbol{m}(A)}$ denotes the derivative of f in the direction $\boldsymbol{m}(A)$ and $\int_0^1(\cdot)\, dt$ is the Riemann integral.

Note that, as a result of Theorem 14.2, the set $\mathcal{R}(\boldsymbol{m})$ is compact and convex and under the assumptions of the theorem above the game $\boldsymbol{v} = f \circ \boldsymbol{m} \in pFNA_{\mathscr{T}}$ because of Proposition 17.4. Hence, the statement above is itself consistent.

The reason for calling (18.6) a *diagonal formula* will become apparent later (see Section 19). For the time being, note that the integration in (18.6) is done along the diagonal of the unit cube $[0,1]^X$.

18.5 The proof of Theorem 18.4 will be carried out in several steps. Following the structure of the corresponding proof in [AUMANN & SHAPLEY 1974], each step is organized as a self-contained lemma, showing which generalizations and extensions of the classical arguments are made.

Lemma. *Let Q be a closed symmetric linear subspace of $FBV_{\mathscr{T}}$. Let $\mathscr{F}$ be the set of functions $f : [0,1] \longrightarrow \mathbb{R}$ of bounded variation such that $f(0) = 0$ and $f(1) = 1$. Let $FBV^1_{\mathscr{T}}$ be the set of all nonnegative nonatomic finite T_∞-measures $\boldsymbol{m}$ such that $\boldsymbol{m}(X) = 1$. If $f \in \mathscr{F}$ and $\boldsymbol{m} \in FNA^1_{\mathscr{T}}$, then $f \circ \boldsymbol{m}$ exists and is contained in $FBV_{\mathscr{T}}$. If the subspace Q is spanned by the family of games $f \circ \boldsymbol{m}$ with $f \in \mathscr{F}$ and $\boldsymbol{m} \in FNA^1_{\mathscr{T}}$ and if a linear operator $\varphi : Q \longrightarrow FBA_{\mathscr{T}}$ satisfies $\|\varphi\| \leq 1$ and $\varphi(f \circ \boldsymbol{m}) = \boldsymbol{m}$ for all $f \in \mathscr{F}$ and $\boldsymbol{m} \in FNA^1_{\mathscr{T}}$, then φ is an Aumann–Shapley value on Q.*

PROOF: Let $f \in \mathscr{F}$ and $\boldsymbol{m} \in FNA^1_{\mathscr{T}}$. Then $\mathcal{R}(\boldsymbol{m}) = [0,1]$ (cf. Corollary 13.14) and $\boldsymbol{v} = f \circ \boldsymbol{m}$ exists as a consequence. Now we show that $\boldsymbol{v}$ has bounded variation. Let $\mathscr{B} = (A_0, A_1, \ldots, A_k)$ be a chain of fuzzy coalitions. Putting $x_i = \boldsymbol{m}(A_i)$ for $i \in \{1, 2, \ldots, k\}$ we have

$$\|\boldsymbol{v}\|_{\mathscr{B}} = \sum_{i=1}^{k} |f(x_i) - f(x_{i-1})|,$$

and this shows that $\|\boldsymbol{v}\|$ is bounded by the total variation of f. Let Q be the closed subspace of FBV spanned by all games $f \circ \boldsymbol{m}$ such that $f \in \mathscr{F}$ and $\boldsymbol{m} \in FNA^1_{\mathscr{T}}$. If π is a permutation of $(X, \mathscr{T})$ and $\boldsymbol{m} \in FNA^1_{\mathscr{T}}$, then $\pi_* \boldsymbol{m} \in FNA^1_{\mathscr{T}}$. Thus for each game $\boldsymbol{v} = f \circ \boldsymbol{m}$ with $f \in \mathscr{F}$ and $\boldsymbol{m} \in FNA^1_{\mathscr{T}}$ we have

$$\varphi(\pi_* \boldsymbol{v}) = \varphi(f \circ \pi_* \boldsymbol{m}) = \pi_* \boldsymbol{m} = \pi_*(\varphi \boldsymbol{v}).$$

Since for each chain of fuzzy coalitions $\mathscr{B} = (A_0, A_1, \ldots, A_k)$ and for each permutation ρ of $(X, \mathscr{T})$ the family of fuzzy coalitions $\mathscr{B}' = (\rho A_i)_{i \in \{1,2,\ldots,k\}}$ is also a chain of fuzzy coalitions, we have

$$\|\pi_* \boldsymbol{v}\|_{\mathscr{B}} = \|\boldsymbol{v}\|_{\mathscr{B}'} \leq \|\boldsymbol{v}\|.$$

This shows that for each permutation η of $(X, \mathscr{T})$ the operator η_* is continuous on $FBV_{\mathscr{T}}$ and, consequently, on Q. Also, the operator φ is continuous on Q since it is bounded. Therefore, for each permutation η of $(X, \mathscr{T})$ the operator $\omega : Q \longrightarrow FBV$ defined by

$$\omega = \varphi \circ \eta_* - \eta_* \circ \varphi$$

is continuous. Since ω vanishes on the spanning subset of Q, it follows that ω is identically zero on Q, and this implies that η satisfies (18.4). The operator $\chi : Q \longrightarrow \mathbb{R}$ given by

$$\chi(\boldsymbol{v}) = \varphi \boldsymbol{v}(X) - \boldsymbol{v}(X)$$

is also linear and continuous. It vanishes on the spanning subset of Q because, by hypothesis, $\varphi \boldsymbol{v}(X) = \boldsymbol{v}(X) = 1$ for each $\boldsymbol{v} = f \circ \boldsymbol{m}$ with $f \in \mathscr{F}$ and $\boldsymbol{m} \in FNA^1_{\mathscr{T}}$. As a consequence, χ is identically zero on Q, and this implies (18.5). It remains to show that φ is positive, i.e., $\varphi\boldsymbol{v}(A) \geq 0$ whenever $\boldsymbol{v} \in Q^+$ and $A \in \mathscr{T}$. Let $\boldsymbol{v} \in Q^+$ and assume that there is a fuzzy coalition A such that $\varphi\boldsymbol{v}(A) < 0$. Then, taking into account $\|\varphi\| \leq 1$, we get

$$\begin{aligned}\boldsymbol{v}(X) &= \|\boldsymbol{v}\| \\ &\geq \|\varphi\boldsymbol{v}\| \\ &\geq |\varphi\boldsymbol{v}(X) - \varphi\boldsymbol{v}(A)| + |\varphi\boldsymbol{v}(A)| \\ &\geq |\boldsymbol{v}(X) - \varphi\boldsymbol{v}(A)| \\ &= \boldsymbol{v}(X) - \varphi\boldsymbol{v}(A) \\ &> \boldsymbol{v}(X),\end{aligned}$$

and this is a contradiction. □

18.6 Lemma. *If $\boldsymbol{m} = (\boldsymbol{m}_1, \ldots, \boldsymbol{m}_n)$ is a nonatomic n-vector T_∞-measure and $f : \mathcal{R}(\boldsymbol{m}) \longrightarrow \mathbb{R}$ is a continuously differentiable function such that $f(0) = 0$, then the function $\boldsymbol{q} : \mathscr{T} \longrightarrow [-\infty, +\infty]$ defined by*

$$\boldsymbol{q}(A) = \int_0^1 f_{\boldsymbol{m}(A)}(t \cdot \boldsymbol{m}(X))\,\mathrm{d}t \tag{18.7}$$

is a finite T_∞-measure.

PROOF: It is sufficient to show that $\boldsymbol{q}$ is a linear combination of finite T_∞-measures. First, assume that the dimension of $\mathcal{R}(\boldsymbol{m})$ equals n. Then its interior $\mathrm{Int}(\mathcal{R}(\boldsymbol{m})) \neq 0$ and

$$f_{\boldsymbol{m}(A)}(x) = \sum_{i=1}^{n} \frac{\partial f}{\partial x_i}(x) \cdot \boldsymbol{m}_i(A) \tag{18.8}$$

for all $A \in \mathscr{T}$ and $x \in \mathcal{R}(\boldsymbol{m})$. Hence, for each $A \in \mathscr{T}$ we have

$$\boldsymbol{q}(A) = \sum_{i=1}^{n} \boldsymbol{m}_i(A) \cdot \int_0^1 \frac{\partial f}{\partial x_i}(t \cdot \boldsymbol{m}_i(X))\,\mathrm{d}t,$$

and Lemma 18.6 is proved in this case. Now assume that the affine dimension of $\mathcal{R}(\boldsymbol{m})$ equals $k < n$. Then there exists a linear mapping $h : \mathbb{R}^n \longrightarrow \mathbb{R}^k$ such that the restriction $\tilde{h} = h/_{\mathcal{R}(\boldsymbol{m})}$ is one-to-one and onto. Put $\boldsymbol{u} = \tilde{h} \circ \boldsymbol{m}$ and $g = f \circ \tilde{h}^{-1}$. Clearly, g is continuously differentiable and $\boldsymbol{u}$ is a nonatomic k-vector T_∞-measure with $\mathcal{R}(\boldsymbol{u})$ having full dimension k within $\mathbb{R}^k$. Thus, formula (18.8) can be applied (with g instead of f) to deduce that

$$\int_0^1 g_{\boldsymbol{u}(A)}(t \cdot \boldsymbol{u}(X))\,\mathrm{d}t$$

is a linear combination of finite T_∞-measures, and, since for all $A \in \mathscr{T}$

$$\boldsymbol{q}(A) = \int_0^1 g_{\boldsymbol{u}(A)}(t \cdot \boldsymbol{u}(X))\, dt,$$

this completes the proof. □

18.7 Lemma. *If $\boldsymbol{m}, f$ and $\boldsymbol{q}$ are as in Lemma 18.6, then*

$$\|\boldsymbol{q}\| \leq \|f \circ \boldsymbol{m}\|. \tag{18.9}$$

PROOF: Let (A^+, A^-) be a Hahn decomposition of X with respect to $\boldsymbol{m}$. Then

$$\|\boldsymbol{q}\| = \boldsymbol{q}^+(X) + \boldsymbol{q}^-(X) = \boldsymbol{q}(A^+) + \boldsymbol{q}(A^-)$$

because of Corollary 9.5 and Theorem 6.3. According to Proposition 12.4, for $k \in \mathbb{N}$ two chains $(B_0^+, B_1^+, \ldots, B_k^+)$ and $(B_0^-, B_1^-, \ldots, B_k^-)$ of fuzzy coalitions can be found such that

$$\emptyset = B_0^+ \leq B_1^+ \leq \ldots \leq B_k^+ = A^+,$$
$$\emptyset = B_0^- \leq B_1^- \leq \ldots \leq B_k^- = A^-,$$

and such that, for each $i \in \{1, 2, \ldots, k\}$ we have

$$\boldsymbol{m}(B_i^+) = \frac{i}{k} \cdot \boldsymbol{m}(A^+),$$
$$\boldsymbol{m}(B_i^-) = \frac{i}{k} \cdot \boldsymbol{m}(A^-).$$

Define $A_0^- = A_0^+ = \emptyset$ and, for $i > 1$,

$$A_i^+ = B_i^+ \ominus B_{i-1}^+,$$
$$A_i^- = B_i^- \ominus B_{i-1}^-.$$

For all $i \in \{1, 2, \ldots, k\}$ we have

$$\boldsymbol{m}(A_i^+) = \frac{1}{k} \cdot \boldsymbol{m}(A^+),$$
$$\boldsymbol{m}(A_i^-) = \frac{1}{k} \cdot \boldsymbol{m}(A^-).$$

It is easily seen that $(A_1^+, A_2^+, \ldots, A_k^+)$ is a T_∞-partition of A^+ and $(A_1^-, A_2^-, \ldots, A_k^-)$ is a T_∞-partition of A^-. Using them, a chain of fuzzy coalitions $\mathscr{B} = (A_0, A_1, \ldots, A_{2k+1})$ can be constructed as follows:

$$A_{2j} = \mathop{\mathbf{S}_\infty}_{i=1}^{j} (A_i^+ \,\mathbf{S}_\infty\, A_i^-),$$
$$A_{2j+1} = A_{2j} \,\mathbf{S}_\infty\, A_{2j+1}^+.$$

Put $y = \boldsymbol{m}(A^+)$, $b = \boldsymbol{m}(X)$ and, consequently, $\boldsymbol{m}(A^-) = b - y$. Therefore, we have

$$\boldsymbol{m}(A_{2j}) = \sum_{i=1}^{j} (A_i^+ \,\mathbf{S}_\infty\, A_i^-) = \sum_{i=1}^{j} (\boldsymbol{m}(A_i^+) + \boldsymbol{m}(A_i^+)) = \frac{j}{k} \cdot \boldsymbol{m}(X)$$

and

$$\boldsymbol{m}(A_{2j+1}) = \boldsymbol{m}(A_{2j}) + \boldsymbol{m}(A_{2j+1}^+) = \frac{1}{k} \cdot (j \cdot \boldsymbol{m}(X) + y)$$

for each $j \in \{1, \dots, k\}$. Taking into account the definition of $\|f \circ \boldsymbol{m}\|$ we get:

$$\begin{aligned}
\|f \circ \boldsymbol{m}\| &\geq \|f \circ \boldsymbol{m}\|_{\mathscr{B}} \\
&= \sum_{j=1}^{2k} |f(\boldsymbol{m}(A_j)) - f(\boldsymbol{m}(A_{j-1}))| \\
&= \sum_{j=0}^{k-1} \left| f\left(\tfrac{j \cdot b + y}{k}\right) - f\left(\tfrac{j \cdot b}{k}\right) \right| + \sum_{j=0}^{k-1} \left| f\left(\tfrac{(j+1) \cdot b}{k}\right) - f\left(\tfrac{j \cdot b + y}{k}\right) \right| \\
&\geq \left| \sum_{j=0}^{k-1} \left[f\left(\tfrac{j \cdot b + y}{k}\right) - f\left(\tfrac{j \cdot b}{k}\right) \right] \right| \\
&\quad + \left| \sum_{j=0}^{k-1} \left[f\left(\tfrac{(j+1) \cdot b}{k}\right) - f\left(\tfrac{j \cdot b + y}{k}\right) \right] \right| .
\end{aligned} \tag{18.10}$$

Since the function $g : \mathcal{R}(\boldsymbol{m}) \longrightarrow \mathbb{R}$ given by

$$g(t) = f\left(\tfrac{j \cdot b}{k} + y \cdot t\right)$$

is continuously differentiable on the compact convex set $\mathscr{R}(\boldsymbol{m})$, it follows that f_y is uniformly continuous on $\mathcal{R}(\boldsymbol{m})$. Thus for a sufficiently large k, formula (18.10) can be rewritten as

$$f\left(\tfrac{j \cdot b + y}{k}\right) - f\left(\tfrac{j \cdot b}{k}\right) = \frac{1}{k} \cdot f_y\left(\tfrac{j \cdot b}{k}\right) + o\left(\tfrac{1}{k}\right), \tag{18.11}$$

where $o : \mathbb{R} \longrightarrow \mathbb{R}$ is a function with $\lim_{t \to 0} o(t) = 0$. Similarly, we obtain

$$f\left(\tfrac{(j+1) \cdot b}{k}\right) - f\left(\tfrac{j \cdot b + y}{k}\right) = \frac{1}{k} \cdot f_{b-y}\left(\tfrac{j \cdot b}{k}\right) + o\left(\tfrac{1}{k}\right). \tag{18.12}$$

Combining (18.10), (18.11) and (18.12), it follows that

$$\|f \circ \boldsymbol{m}\| \geq \left| \sum_{j=0}^{k-1} \frac{1}{k} \cdot f_y\left(\tfrac{j \cdot b}{k}\right) \right| + \left| \sum_{j=0}^{k-1} \frac{1}{k} \cdot f_{b-y}\left(\tfrac{j \cdot b}{k}\right) \right| + o\left(\tfrac{1}{k}\right). \tag{18.13}$$

The sums involved in (18.13) are approximating sums for the two Riemann integrals

$$\int_0^1 f_y(t \cdot b)\,dt$$

and

$$\int_0^1 f_{b-y}(t \cdot b)\,dt,$$

respectively. Therefore, taking the limit $k \to \infty$ in (18.13), we have

$$\|f \circ \boldsymbol{m}\| \geq \left|\int_0^1 f_y(t \cdot b)\,dt\right| + \left|\int_0^1 f_{b-y}(t \cdot b)\,dt\right| = |\boldsymbol{q}(A^+)| + |\boldsymbol{q}(A^-)| = \|\boldsymbol{q}\|,$$

thus completing the proof. □

18.8 The following result will complete the proof of Theorem 18.4:

Lemma. *There exists an Aumann–Shapley value φ on $pFNA_{\mathcal{T}}$ which satisfies equation (18.6).*

PROOF: Let Q be the set of all games $\boldsymbol{v} = f \circ \boldsymbol{m}$ where $\boldsymbol{m}$ is a nonatomic vector T_∞-measure on $\mathcal{T}$ and $f : \mathcal{R}(\boldsymbol{m}) \longrightarrow \mathbb{R}$ is a continuously differentiable function such that $f(0) = 0$. Note that Q is a dense subspace of $pFNA_{\mathcal{T}}$. For $\boldsymbol{v} = f \circ \boldsymbol{m} \in Q$ and $A \in \mathcal{T}$ denote

$$\varphi(f \circ \boldsymbol{m})(A) = \int_0^1 f_{\boldsymbol{m}(A)}(t \cdot \boldsymbol{m}(X))\,dt.$$

We claim that the correspondence $\boldsymbol{v} \longmapsto \varphi\boldsymbol{v}$ is a well-defined continuous linear operator from Q to $FBA_{\mathcal{T}}$. In order to prove that, assume $f \circ \boldsymbol{m} = g \circ \boldsymbol{m}'$ for some nonatomic vector T_∞-measures $\boldsymbol{m}$ and $\boldsymbol{m}'$ and continuously differentiable functions $f : \mathcal{R}(\boldsymbol{m}) \longrightarrow \mathbb{R}$, $g : \mathcal{R}(\boldsymbol{m}') \longrightarrow \mathbb{R}$ satisfying $f(0) = g(0) = 0$. Then $\boldsymbol{m}'' = (\boldsymbol{m}, \boldsymbol{m}')$ is a nonatomic vector T_∞-measure, and the function $h : \mathcal{R}(\boldsymbol{m}) \times \mathcal{R}(\boldsymbol{m}') \longrightarrow \mathbb{R}$ defined by

$$h(x, y) = f(x) - g(y)$$

is continuously differentiable on $\mathcal{R}(\boldsymbol{m}'')$ and satisfies $h(0,0) = 0$. Clearly,

$$\varphi(f \circ \boldsymbol{m}) - \varphi(g \circ \boldsymbol{m}') = \varphi(h \circ \boldsymbol{m}'').$$

By Lemma 18.7 we have

$$\|\varphi(h \circ \boldsymbol{m}'')\| \leq \|h \circ \boldsymbol{m}''\|.$$

Since

$$h \circ \boldsymbol{m}'' = f \circ \boldsymbol{m} - g \circ \boldsymbol{m}' = 0,$$

it follows that $\|h \circ \boldsymbol{m}''\| = 0$ and, implicitly, $\varphi(h \circ \boldsymbol{m}'') = 0$, i.e., $\varphi(f \circ \boldsymbol{m}) = \varphi(g \circ \boldsymbol{m}')$. Linearity of the Riemann integral implies that φ is a linear operator, and $\|\varphi\boldsymbol{v}\| \leq \|\boldsymbol{v}\|$

for all $\boldsymbol{v} \in Q$ (cf. Lemma 18.7) shows that φ is continuous on Q. Also, according to Lemma 18.6, $\varphi\boldsymbol{v} \in FBA_{\mathcal{T}}$ for all $\boldsymbol{v} \in Q$. Hence, φ is a continuous linear operator from Q to $FBA_{\mathcal{T}}$. Since $FBA_{\mathcal{T}}$ is a closed subset of the Banach space $FBV_{\mathcal{T}}$, it follows that $FBA_{\mathcal{T}}$ is complete. Hence, there exists a unique continous linear operator from $pFNA_{\mathcal{T}}$ to $FBA_{\mathcal{T}}$ which extends φ to $pFNA_{\mathcal{T}} = \mathrm{Cl}(Q)$ and has the same norm as φ. This extended operator will again be denoted φ. Lemma 18.5 can be applied in order to show that φ is an Aumann–Shapley value on $pFNA_{\mathcal{T}}$. To this end, note that $pFNA_{\mathcal{T}}$ is a Banach subalgebra of $FBV_{\mathcal{T}}$ which is spanned by all games of the form $f \circ \boldsymbol{m}$ where $\boldsymbol{m} \in FNA^1_{\mathcal{T}}$ and f is a polynomial of the form $f(x) = x^k$ for some $k \in \mathbb{N}$. Since $\|\varphi\boldsymbol{v}\| \le \|\boldsymbol{v}\|$ for all $\boldsymbol{v} = f \circ \boldsymbol{m}$ (f and $\boldsymbol{m}$ as above; see Lemma 18.7), it follows that $\|\varphi\| \le 1$. For $\boldsymbol{v} = \boldsymbol{m}^k$ and for all $A \in \mathcal{T}$

$$\varphi\boldsymbol{v}(A) = \boldsymbol{m}(A) \cdot \int_0^1 \frac{\mathrm{d}}{\mathrm{d}t}(t^k)\,\mathrm{d}t = \boldsymbol{m}(A)$$

i.e., $\varphi\boldsymbol{v} = \boldsymbol{m}$. Consequently, Lemma 18.5 implies that φ is an Aumann–Shapley value on $pFNA_{\mathcal{T}}$. Also, since $\varphi\boldsymbol{m} = \boldsymbol{m}$ for each T_∞-measure of norm 1, it follows that $\|\varphi\| = 1$. □

18.9 Remark. Theorem 18.4 shows that there exists an Aumann–Shapley value on $pFNA_{\mathcal{T}}$. It is a natural question whether this Aumann–Shapley value is unique. AUMANN & SHAPLEY [1974, Chapter 8] have shown that, in general, the answer to this question is negative. However, they have proved that the uniqueness of the Aumann–Shapley value on $pFNA_{\mathcal{T}}$ is guaranteed whenever $\mathcal{T}$ is a σ-algebra of crisp subsets of X which satisfies the so called *standardness assumption* (see [AUMANN & SHAPLEY 1974, Statement (2.1)]). The standardness assumption for a σ-algebra $\mathscr{A}$ of crisp subsets of X requires that the measure space $(X, \mathscr{A})$ be isomorphic to the measure space $([0,1], \mathscr{B})$, where, as usual, $\mathscr{B}$ is the σ-algebra of all Borel subsets of $[0,1]$. The important contribution of the standardness assumption is the existence of a nonatomic measure on $\mathscr{A}$ (namely, the transformation of the Lebesgue measure on $\mathscr{B}$ through the isomorphism existing as a result of the standardness assumption) with respect to which all other nonatomic measures on $\mathscr{A}$ are absolutely continuous. Once this dominant nonatomic measure on $\mathscr{A}$ exists, the uniqueness of the Aumann–Shapley value on $pFNA_{\mathcal{T}}$ is guaranteed. It is interesting to note that if $\mathcal{T}$ is generated and if its underlying σ-algebra $\mathscr{A} = \mathcal{T}^{\vee}$ satisfies the standardness assumption, then the uniqueness proof of the Aumann–Shapley value on $\mathrm{pFNA}_{\mathscr{A}}$ of AUMANN & SHAPLEY [1974, p. 47 ff.] can be carried over without any major modification, showing the uniqueness of the Aumann–Shapley value on $pFNA_{\mathcal{T}}$.

18.10 Note that the Aumann–Shapley value φ, whose existence follows from Theorem 18.4, is the unique continuous Aumann–Shapley value on $pFNA_{\mathcal{T}}$ satisfying the diagonal formula since equality (18.6) defines φ on the generators of $pFNA_{\mathcal{T}}$. In the sequel we shall call φ the *diagonal value* on $pFNA_{\mathcal{T}}$. If, in particular, $\mathcal{T}$ is a σ-algebra of crisp subsets of X, the diagonal value on $pFNA_{\mathcal{T}}$ is exactly the value whose existence also results from Theorem B in [AUMANN & SHAPLEY 1974].

19 Classical Values Revisited

19.1 An intriguing aspect of Theorem 18.4 is that it ensures the existence of unique diagonal values on two spaces of games which are strongly interrelated. To be precise, if X is a nonempty set, if $\mathcal{A}$ is a σ-algebra of subsets of X and if $\mathcal{T} = \mathcal{A}^\wedge$ is the T_∞-tribe generated by $\mathcal{A}$, then, according to Theorem 18.4, diagonal values exist on both spaces, on $pFNA_{\mathcal{A}}$ as well as on $pFNA_{\mathcal{T}}$. This observation leads to the question of whether these Aumann–Shapley values are equivalent (that is, whether one can be obtained from the other by a linear invertible transformation). Note that $pFNA_{\mathcal{A}}$ is exactly the space pNA (with respect to $\mathcal{A}$) on which the existence of the diagonal value also follows from Theorem B of AUMANN & SHAPLEY [1974].

In order to clarify the connection between the diagonal values on $pNA(= pFNA_{\mathcal{A}})$ which we denote ψ and the diagonal value on $pFNA_{\mathcal{T}}$ which we continue to denote φ, we recall that AUMANN & SHAPLEY [1974, Theorem G] have shown the existence of a unique continuous positive linear embedding $\mathrm{e} : pNA \longrightarrow pFNA_{\mathcal{T}}, \ \boldsymbol{v} \longmapsto \mathrm{e}(\boldsymbol{v})$ which is of norm 1, satisfies the condition

$$\mathrm{e}(\boldsymbol{v}_1 \cdot \boldsymbol{v}_2) = \mathrm{e}(\boldsymbol{v}_1) \cdot \mathrm{e}(\boldsymbol{v}_2), \tag{19.1}$$

and gives for each finite nonatomic measure $\boldsymbol{m}$ on $\mathcal{A}$

$$(\mathrm{e}(\boldsymbol{m}))(A) = \int_X A(x)\,\mathrm{d}\boldsymbol{m}(x). \tag{19.2}$$

This operator e is called the *extension operator* and $\mathrm{e}(\boldsymbol{v})$ is called the *extension* of $\boldsymbol{v}$. The image of pNA through e is the closed symmetric subspace of $pFNA_{\mathcal{T}}$ being spanned by the positive integer powers of those finite T_∞-measures on $\mathcal{T}$ whose restrictions to $\mathcal{A}$ are nonatomic; this subspace will be denoted $pFNA^*_{\mathcal{T}}$. By φ^* we shall denote the restriction of the value φ to $pFNA^*_{\mathcal{T}}$.

19.2 Proposition. *If $\mathcal{A}$ is a σ-algebra of crisp subsets of X and if $\mathcal{T} = \mathcal{A}^\wedge$ is the T_∞-tribe generated by $\mathcal{A}$, then the diagonal value ψ on pNA $(= pFNA_{\mathcal{A}})$ is equivalent to the restriction φ^* of the diagonal value φ on $pFNA_{\mathcal{T}}$ to $pFNA^*_{\mathcal{T}}$, i.e., for each $\boldsymbol{v} \in pNA$ and for each $A \in \mathcal{A}$ we have*

$$\psi(\boldsymbol{v})(A) = [(\varphi^* \circ \mathrm{e})^\vee(\boldsymbol{v})](A). \tag{19.3}$$

PROOF: Note that the operator $(\varphi^* \circ e)^\vee$, which assigns to each game $\boldsymbol{v} \in pNA$ the restriction of the T_∞-additive function $\varphi^* \circ e$ to $\mathscr{A}$, is an Aumann–Shapley value on pNA and, according to Theorem 18.4, $(\varphi^* \circ e)^\vee$ satisfies the diagonal formula. Since ψ is the unique Aumann–Shapley value on pNA which satisfies the diagonal formula, we get $\psi = (\varphi^* \circ e)^\vee$, and (19.3) holds. Since e is one–to–one, e has an inverse $e^{-1} : pFNA^*_{\mathscr{A}} \longrightarrow pNA$, and this implies that φ^* is equivalent to ψ. □

19.3 Proposition 19.2 shows that, in general, the *diagonal value on $pFNA_{\mathscr{T}}$* is a proper extension of the diagonal value on pNA. We shall show later (see Chapter V) that for each $\boldsymbol{v} \in pFNA^*_{\mathscr{A}}$, the finitely additive function $\psi(\check{\boldsymbol{v}})$ (where $\check{\boldsymbol{v}}$ is the restriction of $\boldsymbol{v}$ to $\mathscr{A}$) is a measure on $\mathscr{A}$, and thus, whenever $\boldsymbol{v} \in$ p$FNA^*_{\mathscr{A}}$ and $A \in \mathscr{T}$, we have the representation

$$\varphi^*(\boldsymbol{v})(A) = \int_X A(x)\,\mathrm{d}(\psi\,\check{\boldsymbol{v}})(x). \tag{19.4}$$

19.4 The fact emphasized above that the diagonal value on $pFNA_{\mathscr{T}}$ is a proper extension of the diagonal value on pNA becomes obvious if the set of players X is finite, and if $\mathscr{A}$ is the σ-algebra of all crisp subsets of X. In such a case, no nontrivial nonatomic measure on $\mathscr{A}$ exists, i.e., $pNA = \{\mathbf{0}\}$. However, even in this situation and if $\mathscr{T} = \mathscr{A}^\wedge$, the space $pFNA_{\mathscr{T}}$ is nontrivial since the finite nonatomic T_∞-measures $\boldsymbol{m}$ on $\mathscr{T}$ have the form

$$\boldsymbol{m}(A) = \sum_{x \in X} A(x) \cdot m_x, \tag{19.5}$$

for some family $(m_x)_{x\in X} \in [0,1]^X$. The diagonal value φ on $pFNA_{\mathscr{T}}$ in this case is, practically speaking, the well-known *Shapley value* (SHAPLEY [1953]). In order to make this statement more precise, note that due to the finiteness of X, $FBV_{\mathscr{A}}$ equals $GAMES_{\mathscr{A}}$, the linear space of all games on $\mathscr{A}$.

OWEN [1972] has shown that, when X is finite, for each game $\boldsymbol{v} \in FBV_{\mathscr{A}}$ there exists a unique multilinear game $\bar{\boldsymbol{v}}$ on $\mathscr{T}$ such that $\bar{\boldsymbol{v}}(S) = \boldsymbol{v}(S)$ for all $S \in \mathscr{A}$. The game $\bar{\boldsymbol{v}}$, called *multilinear extension of $\boldsymbol{v}$*, is defined as follows:

$$\bar{\boldsymbol{v}}(A) = \sum_{S\in\mathscr{A}} \prod_{x\in S} A(x) \cdot \alpha_S(\boldsymbol{v}),$$

where for each $S \in \mathscr{A}$ we have ($|S|$ denotes the cardinality of the crisp set S)

$$\alpha_S(\boldsymbol{v}) = \sum_{\substack{T\in\mathscr{A} \\ T\subseteq S}} (-1)^{|S|-|T|} \cdot \boldsymbol{v}(T). \tag{19.6}$$

OWEN [1972] has noted that $\bar{\boldsymbol{v}}$ can also be written as

$$\bar{\boldsymbol{v}} = f_{\boldsymbol{v}} \circ \boldsymbol{m} \tag{19.7}$$

where $\boldsymbol{m}$ is the $|X|$-vector T_∞-measure on $\mathscr{T}$ whose coordinates are defined by $\boldsymbol{m}_x(A) = A(x)$ and $f_{\boldsymbol{v}} : [0,1]^X \longrightarrow \mathbb{R}$ is the function given by

$$f_{\boldsymbol{v}}(q) = \sum_{S \in \mathscr{A}} \prod_{x \in S} q_x \cdot \alpha_S(\boldsymbol{v}). \tag{19.8}$$

Since $f_{\boldsymbol{v}}$ is continuously differentiable on $[0,1]^X$ and (19.7) holds, Theorem 17.4 implies that, whenever $\boldsymbol{v} \in FBV_{\mathscr{A}}$, its multilinear extension $\bar{\boldsymbol{v}}$ belongs to $pFNA_{\mathscr{T}}$.

19.5 With these facts in mind, the following statement makes sense.

Proposition. [OWEN 1972]. *If X is a finite set and if $\mathscr{A}$ consists of all crisp subsets of X, then the operator $\chi : FBV_{\mathscr{A}} \longrightarrow FBA_{\mathscr{T}}$ defined by*

$$\chi\boldsymbol{v}(S) = \varphi\bar{\boldsymbol{v}}(S), \tag{19.9}$$

where φ is the diagonal value on $pFNA_{\mathscr{T}}$, is the unique Aumann–Shapley value on $FBV_{\mathscr{A}}$ such that for each $S \in \mathscr{A}$ we have

$$\chi\boldsymbol{v}(S) = \sum_{x \in S} \int_0^1 \frac{\partial f_y}{\partial q_x}(t,t,\ldots,t)\,dt. \tag{19.10}$$

The vector $(\chi_x\boldsymbol{v})_{x\in X} \in \mathbb{R}^X$ with coordinates $\chi_x\boldsymbol{v} = \chi\boldsymbol{v}(\{x\})$ is exactly the Shapley value of $\boldsymbol{v}$, i.e.,

$$\chi_x\boldsymbol{v} = \sum_{\substack{S\in\mathscr{A}\\ x\in S}} \frac{|S|!(|X|-|S|-1)!}{|X|!}[\boldsymbol{v}(S) - \boldsymbol{v}(S\backslash\{x\})]. \tag{19.11}$$

PROOF: We have noted above that the diagonal value φ is the unique Aumann–Shapley value on $pFNA_{\mathscr{T}}$ satisfying (18.6). Taking into account (19.9) and (18.6) we obtain (19.10). It is obvious that χ is an Aumann–Shapley value on $FBV_{\mathscr{A}}$. Observing that for each $q \in [0,1]^X$

$$\frac{\partial f_y}{\partial q_z}(q) = \sum_{\substack{S\in\mathscr{A}\\ z\notin S}} \left[\prod_{x\in S} q_x \cdot \prod_{s\notin S\cup\{z\}} (1-q_s)\cdot[\boldsymbol{v}(S\cup\{z\}) - \boldsymbol{v}(S)]\right],$$

we obtain for each $t \in [0,1]$

$$\frac{\partial f_y}{\partial q_z}(t,t,\ldots,t) = \sum_{\substack{S\in\mathscr{A}\\ z\notin S}} t^{|S|}\cdot(1-t)^{|X|-|S|-1}\cdot[\boldsymbol{v}(S\cup\{z\}) - \boldsymbol{v}(S)].$$

Inserting this expression into (19.10) we obtain equation (19.11), thus completing the proof. □

19.6 Proposition 19.5 shows how the Shapley value can be derived from the diagonal value on $pFNA_{\mathcal{J}}$. AUMANN & SHAPLEY [1974] have shown that it is possible to construct Aumann–Shapley values on various subspaces of BV using the so called *asymptotic method* which, in principle, consists of *expanding* the Shapley value through a procedure which is a sort of converse of the one used in the proof of Proposition 19.5 (see also [MONDERER & NEYMANN 1987]).

Chapter V

Extensions of the Diagonal Value

20 Representation of the Diagonal Value on *pFNA* by Fréchet Differentials

20.1 Theorem 18.4 shows that there always exists an Aumann–Shapley value on the space $pFNA_{\mathscr{T}}$, no matter how the space X and the T_∞-tribe $\mathscr{T}$ (generated or not) are chosen. Note that equation (18.6) in this theorem applies only to games of the form $\boldsymbol{v} = f \circ \boldsymbol{m}$ where $\boldsymbol{m}$ is a vector T_∞-measure and $f : \mathcal{R}(\boldsymbol{m}) \longrightarrow \mathbb{R}$ is a continuously differentiable function such that $f(0) = 0$. It is easily seen that, in the case of a generated T_∞-tribe $\mathscr{T}$, for such games equation (18.6) can be rewritten in the form

$$\varphi\boldsymbol{v}(A) = \int_0^1 D\boldsymbol{v}(tX, A)\,\mathrm{d}\lambda(t),$$

where $D\boldsymbol{v}(tX, \cdot)$ is the Fréchet differential of the game $\boldsymbol{v}$ at the point tX of the diagonal of $[0,1]^X$, and the integral is taken with respect to the Lebesgue measure λ.

The first question is whether this representation of the diagonal value can be extended to all games in $pFNA_{\mathscr{T}}$. This will be answered affirmatively in Theorem 20.4 below.

The fact that, according to Theorem 20.4, the diagonal value $\varphi\boldsymbol{v}$ of each $\boldsymbol{v} \in pFNA_{\mathscr{T}}$ can be represented via the Fréchet differentials $D\boldsymbol{v}(tX, \cdot)$ suggests that the operator φ can be continuously extended to the set of all games which are Fréchet differentiable around the points of the diagonal of the unit cube $[0,1]^X$. This naturally leads to the second question discussed in this chapter, i.e., whether such an extension exists and, if so, whether it is still an Aumann–Shapley value. Proposition 23.2 and Theorem 23.6 provide the answer to this question. They show that the diagonal value φ on $pFNA_{\mathscr{T}}$ can be extended to an Aumann–Shapley value on a closed subspace of $FBV_{\mathscr{T}}$ which we shall call $GDIFFE_{\mathscr{T}}$, and which consists of all games in $FBV_{\mathcal{T}}$ being *Gâteaux differentiable* (and, hence, Fréchet differentiable, as follows from Proposition 23.2) almost everywhere on the diagonal of $[0,1]^X$ and satisfying an additional condition called *diagonal efficiency.*

This result can be compared with the main result in [MERTENS 1980] which shows implicitly that there exists an extension of φ to an Aumann–Shapley value on a space of games with fuzzy coalitions we call here $MDIFFE_{\mathcal{T}}$. The spaces $GDIFFE_{\mathcal{T}}$ and $MDIFFE_{\mathcal{T}}$ are neither identical nor do they contain each other as subspaces (see Section 25) in spite of the fact that $pFNA_{\mathcal{T}}$ is a subspace of their intersection.

Also, neither of the spaces $GDIFFE_{\mathcal{T}}$ and $MDIFFE_{\mathcal{T}}$ is maximal with respect to being the domain of an Aumann–Shapley value. As follows from [MERTENS 1988] and from Theorem 25.5 below, a special type of differential operator can be used in order to further extend the Aumann–Shapley values on $GDIFFE_{\mathcal{T}}$ and $MDIFFE_{\mathcal{T}}$ to continuous Aumann–Shapley values satisfying special *diagonal properties* on specific closed subspaces of $pFBV_{\mathcal{T}}$.

On the one hand, the existence of Aumann–Shapley values satisfying diagonal properties on spaces like $GDIFFE_{\mathcal{T}}$ and $MDIFFE_{\mathcal{T}}$ considerably enlarges the set of market games for which transferable–utility competitive equilibria can be determined via the strong relationship between Aumann–Shapley values and competitive payoff densities, as emphasized in [AUMANN & PERLES 1965] — see also Section 27.

On the other hand, the Aumann–Shapley values presented in this chapter will eventually provide tools for further expanding domains of already known Aumann–Shapley values for games with crisp coalitions (see Section 26).

20.2 In this section we define the concept of *Fréchet differentiability on the diagonal* which we shall deal with. This definition was inspired by Theorem H in [AUMANN & SHAPLEY 1974]. It will be shown later that this concept is equivalent to the Gâteaux differentiability introduced in Section 22. Finally, we state Theorem 20.4 whose proof will be given in Section 22 after some preparatory steps in Section 21. Theorem 20.4 will show that if $\mathcal{T}$ is generated, the games in $pFNA_{\mathcal{T}}$ are Fréchet differentiable on the diagonal and the diagonal value $\varphi \boldsymbol{v}$ can be represented using the Fréchet differential of $\boldsymbol{v}$. In fact, Theorem 20.4 shows more: for each $\boldsymbol{v} \in pFNA_{\mathcal{T}}$ the diagonal value $\varphi \boldsymbol{v}$ is a T_∞-measure. This implicitly proves that the connection between the value φ^* on $pFNA^*_{\mathcal{T}}$ and the diagonal value ψ on pNA announced in (19.4) does indeed hold. Theorem 20.4 is a proper generalization of Theorem H in [AUMANN & SHAPLEY 1974] (see 20.5 and 20.6).

20.3 Let X be a nonempty set and let $\mathcal{T}$ be a generated T_∞-tribe on X and denote $\mathcal{A} = \mathcal{T}^\vee$. By Theorems 3.2 and 3.3, we have $\mathcal{T} = \mathcal{A}^\wedge$, i.e., $\mathcal{T}$ consists of all $\mathcal{A}$-measurable fuzzy sets. Denote by $\mathcal{M}$ the set of all functions $f : X \longrightarrow \mathbb{R}$ which are bounded and $\mathcal{A}$-measurable. The set $\mathcal{M}$ is a Banach space with pointwise addition and scalar multiplication induced from $\mathbb{R}$, and with the norm

$$\|f\| = \sup\{|f(x)| \mid x \in X\}. \tag{20.1}$$

According to Theorem 3.2, $\mathcal{T}$ equals the subset of $\mathcal{M}$ consisting of all nonnegative functions in $\mathcal{M}$ of norm less than or equal to 1. As usual, $\mathrm{Int}(\mathcal{T})$ denotes the interior of $\mathcal{T}$ in the Banach space $\mathcal{M}$. Whenever $t \in]0,1[$ then the constant fuzzy set tX defined by $tX(x) = t$ belongs to $\mathrm{Int}(\mathcal{T})$. Also, if $\tau \in]0,1[$, then for each $A \in \mathcal{T}$ the fuzzy set τA is defined by $\tau A(x) = \tau \cdot A(x)$ belongs to $\mathcal{T}$ (since $\mathcal{T}$ is generated) and if $\tau \le 1 - t$, we have

$$tX \, \mathbf{S}_\infty \, \tau A = tX + \tau A.$$

Note that if $f \in \mathcal{M}$ such that $\|f\|$ is sufficiently small, then $tX + f \in \mathcal{T}$ too. In the sequel λ will denote the Lebesgue measure on $[0,1]$.

Definition. A game $\boldsymbol{v} \in pFBV_{\mathcal{T}}$ is called *Fréchet differentiable on the diagonal* (or simply *F-differentiable*) if it satisfies the following two conditions:

(i) The restriction of $\boldsymbol{v}$ to $\mathrm{Int}(\mathcal{T})$ is Fréchet differentiable at tX for λ-almost all $t \in]0,1[$, i.e., for λ-almost all $t \in]0,1[$ there exists a linear operator $D\boldsymbol{v}(tX,\cdot) : \mathcal{M} \longrightarrow \mathbb{R}$ such that

$$\lim_{\|f\| \searrow 0} \frac{\boldsymbol{v}(tX+f) - \boldsymbol{v}(tX) - D\boldsymbol{v}(tX,f)}{\|f\|} = 0; \tag{20.2}$$

(ii) For each $A \in \mathcal{T}$ the function $t \longmapsto D\boldsymbol{v}(tX,A)$ is λ-integrable on $[0,1]$ (in the sense that there exists a λ-integrable function $h_A : [0,1] \longrightarrow \mathbb{R}$ such that $D\boldsymbol{v}(tX,A) = h_A(t)$ whenever $D\boldsymbol{v}(tX,\cdot)$ exists).

Note that we do not require the continuity of the Fréchet differential $D\boldsymbol{v}(tX,\cdot)$. If $\boldsymbol{v}$ is F-differentiable and if for λ-almost all $t \in]0,1[$ the Fréchet differential $D\boldsymbol{v}(tX,\cdot)$ is continuous on $\mathcal{M}$, then the game $\boldsymbol{v}$ is called *continuously F-differentiable* (*on the diagonal*). We shall see in Proposition 23.2 that F-differentiable games in $pFBV_{\mathcal{T}}$ are necessarily continuously F-differentiable. Also, it will be shown later that in the definition above the only necessary condition is (i) since condition (ii) follows from (i) combined with Proposition 23.2.

Observe that the F-differentiability of $\boldsymbol{v}$ is, in fact, a Fréchet differentiability condition almost everywhere on the diagonal of the unit cube $[0,1]^X$ for the restriction of $\boldsymbol{v}$ to the open subset $\mathrm{Int}(\mathcal{T})$. Therefore, the term *differentiable on the diagonal* is a proper name for this concept.

20.4 We sometimes use the abbreviations introduced in Section 19, i.e., *FBV* and *pFNA* stand for $FBV_{\mathcal{T}}$ and $pFNA_{\mathcal{T}}$, respectively, while *BV* and *pNA* stand for $FBV_{\mathcal{A}}$ and $pFNA_{\mathcal{A}}$, respectively. The following result is the basis for further extensions of the diagonal value φ on *pFNA* to larger spaces.

Theorem. *Let $\mathscr{T}$ be a generated T_∞-tribe. Then the games in $pFNA_{\mathscr{T}}$ are continuously F-differentiable, for each $\boldsymbol{v} \in pFNA_{\mathscr{T}}$ the value $\varphi\boldsymbol{v}$ is a T_∞-measure, and for each $A \in \mathscr{T}$ the diagonal value φ on $pFNA_{\mathscr{T}}$ can be represented by the following diagonal formula:*

$$\varphi\boldsymbol{v}(A) = \int_0^1 D\boldsymbol{v}(tX, A)\, \mathrm{d}\lambda(t). \tag{20.3}$$

Note that the diagonal formula (20.3) is in fact a generalization of (18.6) since (20.3) reduces to (18.6) in the case $\boldsymbol{v} = f \circ \boldsymbol{m}$ whenever $\boldsymbol{m}$ and f are as in Theorem 18.4. This justifies calling not only (18.6) but also (20.3) a *diagonal formula.*

The proof of this theorem, being based on some preparatory steps presented in Section 21 will be postponed until Section 22.

20.5 Let ψ be the diagonal value on pNA $(= pFNA_{\mathscr{A}})$. It was pointed out in Section 19.2 that $\psi = (\varphi^* \circ e)^\vee$ where $e : pNA \longrightarrow pFNA^*$ is the extension operator satisfying (19.1) and (19.2) and φ^* is the restriction of φ to $pFNA^*$. Combining this fact with Theorem 20.4 we obtain the following:

Corollary. *If $\boldsymbol{v} \in pNA$, then the extension game $\boldsymbol{v}^* = e(\boldsymbol{v})$ is continuously F-differentiable on the diagonal, for each $S \in \mathscr{A}$ the function $t \longmapsto D\boldsymbol{v}^*(tX, S)$ is λ-integrable on $[0, 1]$ and the diagonal value ψ on pNA can be represented by the following diagonal formula:*

$$\psi\boldsymbol{v}(S) = \int_0^1 D\boldsymbol{v}^*(tX, S)\, \mathrm{d}\lambda(t).$$

20.6 Corollary 20.5 generalizes Theorem H in [AUMANN & SHAPLEY 1974] in the sense that, unlike this Theorem H, it does not require the measure space $(X, \mathscr{A})$ to satisfy the *standardness assumption* (see Section 18.9). The standardness assumption is substantially involved in the proof since Aumann and Shapley use the internality of the space pNA (see AUMANN & SHAPLEY [1974, Proposition 7.19 and p. 154]). A careful analysis of the proof of Theorem G in [AUMANN & SHAPLEY 1974] shows that the existence of the extension operator e is proven without using the standardness assumption. Since in the proof of Theorem 18.4 (which guarantees the existence of ψ and φ) as well as in the proof of Theorem 20.4 (see Sections 21 and 22 below) the standardness assumption is not needed, Corollary 20.5 follows from these results (without the standardness assumption). Note that the standardness assumption is quite restrictive. For instance, games with finitely or countably many players defined on a σ-algebra of crisp subsets of X do not satisfy this condition. Of course, it is possible to obtain Aumann–Shapley values for games with finitely many players via the constructive procedure of SHAPLEY [1953] — see also Proposition 19.5 — and

Aumann–Shapley values for games with countably many players via the limit process introduced by ARTSTEIN [1970]. The elimination of the standardness assumption in Theorem 20.4 as well as in Corollary 20.5 allows us to deduce these results in a unified approach.

21 Dini Derivatives of Games in *FBV*

21.1 In this section we study the behavior of *Dini derivatives* of games in $FBV_{\mathscr{T}}$ at points on the diagonal of the unit cube $[0,1]^X$, where $\mathscr{T}$ is a generated T_∞-tribe on X and $\mathscr{A} = \mathscr{T}^\vee$. In spite of the fact that Dini derivatives of games in $pFNA^*$ were implicitly used by AUMANN & SHAPLEY [1974] in the proof of Theorem H (see also Corollary 20.5 and Section 20.6 above), it was MERTENS [1980] who first fully realized their importance for the theory of Aumann–Shapley values without using the term Dini derivative explicitly. In this section we point out some properties of the Dini derivatives of games in $FBV_{\mathscr{T}}$ in a context which is more general than that considered by MERTENS [1980, 1988].

21.2 Let $\boldsymbol{v}$ be a game in $FBV_{\mathscr{T}}$ and $A \in \mathscr{M}$. Define the function $h_{\boldsymbol{v},A} : \mathbb{R}^2 \longrightarrow \mathbb{R}$ by

$$h_{\boldsymbol{v},A}(t,\tau) = \boldsymbol{v}[\max(\emptyset, \min(1, tX + \tau A))]. \tag{21.1}$$

Note that $h_{\boldsymbol{v},A}$ is well-defined and that

$$h_{\boldsymbol{v},A}(t,\tau) = \boldsymbol{v}(tX \,\mathbf{S}_\infty\, \tau A) \tag{21.2}$$

whenever $(t,\tau) \in [0,1]^2$ and $A \in \mathscr{T}$. The *upper* and *lower Dini derivatives of* $h_{\boldsymbol{v},A}$ *at the point* $a \in \mathbb{R}^2$ *in the direction* $d \in \mathbb{R}^2$ are denoted $\delta^+ h_{\boldsymbol{v},A}(a,d)$ and $\delta^- h_{\boldsymbol{v},A}(a,d)$, respectively. Recall that

$$\delta^+ h_{\boldsymbol{v},A}(a,d) = \limsup_{\tau \searrow 0} \frac{h_{\boldsymbol{v},A}(a+\tau d) - h_{\boldsymbol{v},A}(a)}{\tau}, \tag{21.3}$$

$$\delta^- h_{\boldsymbol{v},A}(a,d) = \liminf_{\tau \searrow 0} \frac{h_{\boldsymbol{v},A}(a+\tau d) - h_{\boldsymbol{v},A}(a)}{\tau}. \tag{21.4}$$

Naturally, either one of $\delta^+ h_{\boldsymbol{v},A}(a,d)$ and $\delta^- h_{\boldsymbol{v},A}(a,d)$ or both Dini derivatives may be infinite.

Definition. Let $t \in [0,1]$, $A \in \mathscr{T}$ and $\boldsymbol{v} \in FBV_{\mathscr{T}}$. The upper (lower) Dini derivative of $h_{\boldsymbol{v},A}$ at the point $a = (t,0)$ in the direction $d = (0,1)$ will be called *upper (lower) Dini derivative of the game* $\boldsymbol{v}$ *at* tX *in the direction* A; they will be denoted $\delta^+\boldsymbol{v}(t,A)$ and $\delta^-\boldsymbol{v}(t,A)$, respectively.

According to (21.2), (21.3) and (21.4) we have

$$\delta^+ \boldsymbol{v}(t, A) = \limsup_{\tau \searrow 0} \frac{\boldsymbol{v}(tX \, \mathbf{S}_\infty \, \tau A) - \boldsymbol{v}(tX)}{\tau} \tag{21.5}$$

and

$$\delta^- \boldsymbol{v}(t, A) = \liminf_{\tau \searrow 0} \frac{\boldsymbol{v}(tX \, \mathbf{S}_\infty \, \tau A) - \boldsymbol{v}(tX)}{\tau}. \tag{21.6}$$

21.3 If $\boldsymbol{v} \in FBV^+_{\mathscr{T}}$ and $A \in \mathscr{T}$, then $h_{\boldsymbol{v},A}$ is monotone on $\mathbb{R}^2$, i.e.,

$$h_{\boldsymbol{v},A}(t, \tau) \leq h_{\boldsymbol{v},A}(t', \tau')$$

whenever $t \leq t'$ and $\tau \leq \tau'$. Therefore, $h_{\boldsymbol{v},A}$ is monotone in each direction $d \in \mathbb{R}^2_+$, i.e., for each $a \in \mathbb{R}^2$ the function $\alpha \longmapsto h_{\boldsymbol{v},A}(a + \alpha d)$ is nondecreasing on $\mathbb{R}$. This implies that $h_{\boldsymbol{v},A}$ is lower semicontinuous in each direction $d \in \mathbb{R}^2_+ \setminus \{0\}$: if $a \in \mathbb{R}^2_+$ and if $h_{\boldsymbol{v},A}(a) > r$ then there exists an $\alpha_0 > 0$ such that for all $\alpha \in [0, \alpha_0]$ we have $h_{\boldsymbol{v},A}(a + \alpha d) > r$ (take, for instance, $\alpha_0 = 1$). It follows from Proposition 7 in [CHABRILLAC & CROUZEIX 1987] that the upper and the lower Dini derivatives $\delta^+ h_{\boldsymbol{v},A}(\cdot, d)$ and $\delta^- h_{\boldsymbol{v},A}(\cdot, d)$ for each direction $d \in \mathbb{R}^2_+$, are Borel measurable on $\mathbb{R}^2$. In particular, $\delta^+ h_{\boldsymbol{v},A}(\cdot, d)$ and $\delta^- h_{\boldsymbol{v},A}(\cdot, d)$ are Borel measurable if $d = (0, 1)$. Summarizing the discussion above, we can formulate the following result:

Lemma. *If $\boldsymbol{v} \in FBV^+_{\mathscr{T}}$ and if $A \in \mathscr{T}$ then we have:*

(i) $h_{\boldsymbol{v},A}$ is monotone and lower semicontinuous in each direction $d \in \mathbb{R}^2_+$;

(ii) the Dini derivatives $\delta^+ h_{\boldsymbol{v},A}(\cdot, d)$ and $\delta^- h_{\boldsymbol{v},A}(\cdot, d)$ are Borel measurable on $\mathbb{R}^2_+$ in each direction $d \in \mathbb{R}^2_+$.

21.4 Let $\boldsymbol{v} \in FBV_{\mathscr{T}}$. The function $g_{\boldsymbol{v}} : [0, 1] \longrightarrow \mathbb{R}$ defined by

$$g_{\boldsymbol{v}}(t) = \boldsymbol{v}(tX)$$

will be called the *diagonal function of* $\boldsymbol{v}$. Obviously, $g_{\boldsymbol{v}}$ is nondecreasing whenever $\boldsymbol{v}$ is monotone. Also $g_{\boldsymbol{v}} = g_{\boldsymbol{v}^+} - g_{\boldsymbol{v}^-}$, where $\boldsymbol{v}^+$ and $\boldsymbol{v}^-$ are the upper and the lower variation of $\boldsymbol{v}$, respectively. As a consequence, $g_{\boldsymbol{v}}$ is of bounded variation. With these facts in mind we can prove the following:

Lemma. *If $\boldsymbol{v} \in FBV^+_{\mathscr{T}}$, then for each $A \in \mathscr{T}$ the functions $\delta^+ \boldsymbol{v}(\cdot, A)$ and $\delta^- \boldsymbol{v}(\cdot, A)$ are λ-integrable on $[0, 1]$ and satisfy the inequalities*

$$0 \leq \int_0^1 \delta^- \boldsymbol{v}(t, A) \, \mathrm{d}\lambda(t) \leq \int_0^1 \delta^+ \boldsymbol{v}(t, A) \, \mathrm{d}\lambda(t) \leq \boldsymbol{v}(X). \tag{21.7}$$

PROOF: Note that $\delta^+\boldsymbol{v}(\cdot,A)$ and $\delta^-\boldsymbol{v}(\cdot,A)$ equal the composition of $\delta^+h_{\boldsymbol{v},A}(\cdot,d)$ and $\delta^-h_{\boldsymbol{v},A}(\cdot,d)$, respectively, where $d=(0,1)$, with the continuous embedding of $[0,1]$ into $[0,1]^2$ defined by $t\longmapsto(t,0)$. From Lemma 21.3 it follows that $\delta^+\boldsymbol{v}(\cdot,A)$ and $\delta^-\boldsymbol{v}(\cdot,A)$ are Borel measurable and nonnegative. Hence, the integrals in (21.7) exist and the first and the second inequality in (21.7) hold since $\boldsymbol{v}$ is monotone and $g_{\boldsymbol{v}}$ is nondecreasing on $[0,1]$. Hence, $g_{\boldsymbol{v}}$ is differentiable λ-a.e. on $[0,1]$, its derivative $g'_{\boldsymbol{v}}$ is nonnegative and λ-integrable on $[0,1]$, and we have

$$\int_0^1 g'_{\boldsymbol{v}}(t)\,\mathrm{d}t \le g_{\boldsymbol{v}}(1)-g_{\boldsymbol{v}}(0)=\boldsymbol{v}(X). \tag{21.8}$$

For each $t\in\,]0,1[$ and for each $\tau\in\,]0,1-t[$ we get

$$\begin{aligned} 0 &\le \frac{\boldsymbol{v}(tX\,\mathbf{S}_\infty\,\tau A)-\boldsymbol{v}(tX)}{\tau}\\ &= \frac{\boldsymbol{v}(tX+\tau A)-\boldsymbol{v}(tX)}{\tau}\\ &\le \frac{\boldsymbol{v}((t+\tau)X)-\boldsymbol{v}(tX)}{\tau}\\ &= \frac{g_{\boldsymbol{v}}(t+\tau)-g_{\boldsymbol{v}}(t)}{\tau}. \end{aligned}$$

Taking the upper limit for $\tau\searrow 0$ in these inequalities we obtain

$$\delta^+\boldsymbol{v}(t,A)\le g'_{\boldsymbol{v}}(t) \tag{21.9}$$

whenever $g_{\boldsymbol{v}}$ is differentiable at $t\in\,]0,1[$. Computing the integral over $[0,1]$ of both sides of the inequality (21.9) with respect to λ and using (21.8), the last inequality in (21.7) follows. □

21.5 Lemma 21.4 establishes the λ-integrability of $\delta^-\boldsymbol{v}(\cdot,A)$ and of $\delta^+\boldsymbol{v}(\cdot,A)$ as well as inequalities (21.7) in the case when $\boldsymbol{v}$ is monotone. We would like to have a relation similar to (21.7) for each game $\boldsymbol{v}$ in $FBV_{\mathcal{T}}$. Since, in general, the functions $\delta^-\boldsymbol{v}(t,A)$ and $\delta^+\boldsymbol{v}(t,A)$ are not linear in $\boldsymbol{v}$, it seems difficult (if not impossible) to deduce the Borel measurability of the functions $\delta^-\boldsymbol{v}(\cdot,A)$ or $\delta^+\boldsymbol{v}(\cdot,A)$ by representing $\boldsymbol{v}$ as a difference of two monotone games. However, we can establish a relation somewhat similar to (21.7) for each game $\boldsymbol{v}$ in $FBV_{\mathcal{T}}$ using upper integrals instead of Lebesgue integrals.

Recall that the *upper integral*

$$\overline{\int_0^1} g(t)\,\mathrm{d}\lambda(t)$$

of a function $g : [0,1] \longrightarrow [0,\infty]$ with respect to λ is defined by

$$\overline{\int_0^1} g(t)\,\mathrm{d}\lambda(t) = \inf\left\{\int_0^1 k(t)\,\mathrm{d}\lambda(t)\right\} \tag{21.10}$$

where the infimum is taken over all Borel measurable functions $k : [0,1] \longrightarrow [0,\infty]$ satisfying $k \geq g$ λ-a.e. Obviously, if g is Borel measurable then the upper integral of g and the Lebesgue integral of g with respect to λ coincide. If there is no Borel measurable function $k : [0,1] \longrightarrow [0,\infty]$ such that $k \geq g$ λ-a.e. then, by definition, the value of the upper integral of g equals ∞. Note that the upper integral is monotone with respect to g. Define

$$\partial^+ \boldsymbol{v}(t,A) = \limsup_{\tau\searrow 0} \frac{|\boldsymbol{v}(tX \,\mathbf{S}_\infty\, \tau A) - \boldsymbol{v}(tX)|}{\tau} \tag{21.11}$$

and

$$\Delta^+ \boldsymbol{v}(t) = \limsup_{\|f\|\searrow 0} \frac{|\boldsymbol{v}(tX + f) - \boldsymbol{v}(tX)|}{\|f\|}. \tag{21.12}$$

Lemma. *If $\boldsymbol{v} \in FBV_{\mathscr{T}}$, then for each $A \in \mathscr{T}$*

$$\overline{\int_0^1} \partial^+ \boldsymbol{v}(t,A)\,\mathrm{d}\lambda(t) \leq \overline{\int_0^1} \Delta^+ \boldsymbol{v}(t)\,\mathrm{d}\lambda(t) \leq \|\boldsymbol{v}\|. \tag{21.13}$$

PROOF: Note that $\partial^+ \boldsymbol{v}(t,A) \leq \Delta^+ \boldsymbol{v}(t)$ for each $t \in [0,1]$ implying the first inequality in (21.13). In order to prove the second inequality, assume first that $\boldsymbol{v}$ is monotone. Then $g_{\boldsymbol{v}}$ is nondecreasing and, therefore, λ-a.e. differentiable on $[0,1]$, and its derivative $g'_{\boldsymbol{v}}$ is measurable. If $\|f\|$ is sufficiently small, we have

$$\frac{|\boldsymbol{v}(tX + f) - \boldsymbol{v}(tX)|}{\|f\|} \leq \frac{g_{\boldsymbol{v}}(t + \|f\|) - g_{\boldsymbol{v}}(t)}{\|f\|}$$

Therefore, whenever $g'_{\boldsymbol{v}}(t)$ exists, we get

$$\Delta^+ \boldsymbol{v}(t) \leq g'_{\boldsymbol{v}}(t).$$

Integrating both sides of this inequality and taking into account (21.8) which holds in this case, the last inequality in (21.13) follows for monotone games $\boldsymbol{v}$. Now suppose that $\boldsymbol{v}$ is not monotone. Denote $\boldsymbol{u}$ and $\boldsymbol{w}$ as the upper and the lower variation of $\boldsymbol{v}$, respectively. Note that for each $t \in [0,1]$ the function $\boldsymbol{v} \longmapsto \Delta^+ \boldsymbol{v}(t)$ is subadditive. Hence, for all $t \in [0,1]$ we have

$$\Delta^+ \boldsymbol{v}(t) \leq \Delta^+ \boldsymbol{u}(t) + \Delta^+ \boldsymbol{w}(t).$$

Integrating both sides of this inequality and taking into account that $\boldsymbol{u}$ and $\boldsymbol{w}$ are monotone and that for monotone games (21.13) has already been already proven, we obtain

$$\begin{aligned}\overline{\int_0^1} \Delta^+\boldsymbol{v}(t)\,\mathrm{d}\lambda(t) &\le \overline{\int_0^1} \Delta^+\boldsymbol{u}(t)\,\mathrm{d}\lambda(t) + \overline{\int_0^1} \Delta^+\boldsymbol{w}(t)\,\mathrm{d}\lambda(t)\\ &\le \boldsymbol{u}(X) + \boldsymbol{w}(X)\\ &= \|\boldsymbol{v}\|,\end{aligned}$$

which completes the proof. □

21.6 For each $A \in \mathscr{T}$ define the function*) $\Delta_A : FBV_{\mathscr{T}} \longrightarrow [0,\infty]$, $\boldsymbol{v} \longmapsto \Delta_A\boldsymbol{v}$, by

$$\Delta_A\boldsymbol{v} = \overline{\int_0^1} [\delta^+\boldsymbol{v}(t,A) - \delta^-\boldsymbol{v}(t,A)]\,\mathrm{d}\lambda(t). \tag{21.14}$$

This function is well-defined since the integrand is nonnegative.

Lemma. *For each $A \in \mathscr{T}$ the function Δ_A is a seminorm on $FBV_{\mathscr{T}}$ and, whenever $\boldsymbol{v} \in FBV_{\mathscr{T}}$, we have*

$$\Delta_A \le \|\boldsymbol{v}\|. \tag{21.15}$$

PROOF: It is obvious that Δ_A is nonnegative and positively homogeneous. Since $\delta^+\boldsymbol{v}(t,A)$ is subadditive and $\delta^-\boldsymbol{v}(t,A)$ is superadditive with respect to $\boldsymbol{v}$, it follows that Δ_A is subadditive on $FBV_{\mathscr{T}}$. Note that $\Delta_A(-\boldsymbol{v}) = \Delta_A\boldsymbol{v}$. Denote $\boldsymbol{u}$ and $\boldsymbol{w}$ as the upper and the lower variation of $\boldsymbol{v}$, respectively. We then have

$$\begin{aligned}\Delta_A\boldsymbol{v} &= \Delta_A(\boldsymbol{u} - \boldsymbol{w})\\ &< \Delta_A\boldsymbol{u} + \Delta_A(-\boldsymbol{w})\\ &= \Delta_A\boldsymbol{u} + \Delta_A\boldsymbol{w}\\ &= \overline{\int_0^1} [\delta^+\boldsymbol{u}(t,A) - \delta^-\boldsymbol{u}(t,A)]\,\mathrm{d}\lambda(t) + \overline{\int_0^1} [\delta^+\boldsymbol{w}(t,A) - \delta^-\boldsymbol{w}(t,A)]\,\mathrm{d}\lambda(t).\end{aligned}$$

According to Lemma 21.4, we can replace the upper integrals by Lebesgue integrals with respect to λ since $\boldsymbol{u}$ and $\boldsymbol{w}$ are monotone. Hence,

$$\Delta_A\boldsymbol{v} \le \int_0^1 [\delta^+\boldsymbol{u}(t,A) + \delta^+\boldsymbol{w}(t,A)]\,\mathrm{d}\lambda(t) - \int_0^1 [\delta^-\boldsymbol{u}(t,A) + \delta^-\boldsymbol{w}(t,A)]\,\mathrm{d}\lambda(t).$$

Note that the second integral is nonnegative and finite (cf. (21.7)). Hence, according to Lemma 21.4, we get

$$\Delta_A\boldsymbol{v} \le \int_0^1 [\delta^+\boldsymbol{u}(t,A) + \delta^+\boldsymbol{w}(t,A)]\,\mathrm{d}\lambda(t) \le \boldsymbol{u}(X) + \boldsymbol{w}(X) = \|\boldsymbol{v}\|.$$

□

*)We make the usual convention that $\infty - \infty = 0$

21.7 An important property of the Dini derivatives of games in $FBV_{\mathcal{T}}$ is their special uniform continuity at $A = \emptyset$ which is a consequence of the following result:

Lemma. *If $\boldsymbol{v} \in FBV^{+}_{\mathcal{T}}$, then $\delta^{+}\boldsymbol{v}(t,\cdot)$ and $\delta^{-}\boldsymbol{v}(t,\cdot)$ are monotone on $\mathcal{T}$ and we have*

$$0 \leq \delta^{-}\boldsymbol{v}(t, A) \leq \delta^{+}\boldsymbol{v}(t, A) \leq \|A\| \cdot \delta^{+}\boldsymbol{v}(t, X) \tag{21.16}$$

for each $t \in [0,1]$ and for each $A \in \mathcal{T}$. Moreover, the functions $\delta^{+}\boldsymbol{v}(t,\cdot)$ and $\delta^{-}\boldsymbol{v}(t,\cdot)$ are finite whenever $g_{\boldsymbol{v}}$ is differentiable at t.

PROOF: Note that if $A, B \in \mathcal{T}$ such that $A \leq B$ then for each $t \in]0,1]$ we have

$$\frac{\boldsymbol{v}(tX \, \mathbf{S}_{\infty} \, \tau A) - \boldsymbol{v}(tX)}{\tau} \leq \frac{\boldsymbol{v}(tX \, \mathbf{S}_{\infty} \, \tau B) - \boldsymbol{v}(tX)}{\tau}$$

because of the monotonicity of $\boldsymbol{v}$. This implies that the Dini derivatives $\delta^{+}\boldsymbol{v}(t,\cdot)$ and $\delta^{-}\boldsymbol{v}(t,\cdot)$ are monotone. Monotonicity of $\boldsymbol{v}$ implies

$$\begin{aligned}
\frac{\boldsymbol{v}(tX \, \mathbf{S}_{\infty} \, \tau A) - \boldsymbol{v}(tX)}{\tau} &= \frac{\boldsymbol{v}(tX + \tau A) - \boldsymbol{v}(tX)}{\tau} \\
&\leq \frac{\boldsymbol{v}(tX + \tau\|A\|X) - \boldsymbol{v}(tX)}{\tau} \\
&= \|A\| \cdot \frac{\boldsymbol{v}(tX + \tau\|A\|X) - \boldsymbol{v}(tX)}{\tau\|A\|} \\
&= \|A\| \cdot \frac{g_{\boldsymbol{v}}(t + \tau\|A\|) - g_{\boldsymbol{v}}(t)}{\tau\|A\|}
\end{aligned}$$

whenever $\tau \in]0,1]$ is sufficiently small and $\|A\| \neq 0$. Taking the upper limit $\tau \searrow 0$ in this inequality, we obtain

$$\delta^{+}\boldsymbol{v}(t, A) \leq \|A\| \cdot \delta^{+}\boldsymbol{v}(t, X) = \|A\| \cdot \limsup_{\tau \searrow 0} \frac{g_{\boldsymbol{v}}(t + \tau\|A\|) - g_{\boldsymbol{v}}(t)}{\tau\|A\|} \tag{21.17}$$

and this proves (21.16). Note that if $g_{\boldsymbol{v}}$ is differentiable at the point $t \in [0,1]$, then the upper limit in (21.17) equals $g'_{\boldsymbol{v}}(t)$. Hence, if $g_{\boldsymbol{v}}$ is differentiable at $t \in]0,1[$ then $\delta^{+}\boldsymbol{v}(t,\cdot)$ is finite. □

21.8 Corollary. *For each $\boldsymbol{v} \in FBV_{\mathcal{T}}$, for each $t \in [0,1]$ and for each $A \in \mathcal{T}$ we have*

$$|\delta^{+}\boldsymbol{v}(t, A)| \leq \|A\| \cdot (\delta^{+}\boldsymbol{v}^{+}(t, X) + \delta^{+}\boldsymbol{v}^{-}(t, X)), \tag{21.18}$$

and

$$|\delta^{-}\boldsymbol{v}(t, A)| \leq \|A\| \cdot (\delta^{+}\boldsymbol{v}^{+}(t, X) + \delta^{+}\boldsymbol{v}^{-}(t, X)), \tag{21.19}$$

where $\boldsymbol{v}^{+}$ and $\boldsymbol{v}^{-}$ are the upper and the lower variation of $\boldsymbol{v}$, respectively. For λ-almost all $t \in [0,1]$ both Dini derivatives $\delta^{+}\boldsymbol{v}(t,\cdot)$ and $\delta^{-}\boldsymbol{v}(t,\cdot)$ are finite and uniformly continuous on $\mathcal{T}$ at $A = \emptyset$.

PROOF: Note that for each $A \in \mathscr{T}$ and for each $t \in [0,1]$, we have

$$\begin{aligned}\delta^- \boldsymbol{v}^+(t,A) - \delta^+ \boldsymbol{v}^-(t,A) &\leq \delta^- \boldsymbol{v}(t,A) \\ &\leq \delta^+ \boldsymbol{v}(t,A) \\ &\leq \delta^+ \boldsymbol{v}^+(t,A) - \delta^- \boldsymbol{v}^-(t,A).\end{aligned}$$

Since $\boldsymbol{v}^-$ and $\boldsymbol{v}^+$ are monotone, applying Lemma 21.7 yields

$$\begin{aligned}-\|A\| \cdot \delta^+ \boldsymbol{v}^-(t,X) &\leq -\delta^+ \boldsymbol{v}^-(t,A) \\ &\leq \delta^- \boldsymbol{v}(t,A) \\ &\leq \delta^+ \boldsymbol{v}(t,A) \\ &\leq \delta^+ \boldsymbol{v}^+(t,A) \\ &\leq \|A\| \cdot \delta^+ \boldsymbol{v}(t,A).\end{aligned}$$

Hence,

$$\begin{aligned}|\delta^+ \boldsymbol{v}(t,A)| &\leq \max[\delta^+ \boldsymbol{v}^+(t,A), \delta^+ \boldsymbol{v}^-(t,A)] \\ &\leq \|A\| \cdot \max[\delta^+ \boldsymbol{v}^+(t,X), \delta^+ \boldsymbol{v}^-(t,X)]\end{aligned}$$

and (21.18) follows. The proof of (21.19) is completely analogous. Since by Lemma 21.7, $\delta^+ \boldsymbol{v}^+(t,X)$ and $\delta^+ \boldsymbol{v}^-(t,X)$ are finite for λ-almost all $t \in [0,1]$, $\delta^+ \boldsymbol{v}(t,X)$ and $\delta^- \boldsymbol{v}(t,X)$ are finite for λ-almost all $t \in [0,1]$. Whenever the Dini derivatives $\delta^+ \boldsymbol{v}(t,\cdot)$ and $\delta^- \boldsymbol{v}(t,\cdot)$ are finite, they are uniformly continuous on $\mathscr{T}$ at $A = \emptyset$ due to (21.18), (21.19) and $\delta^+ \boldsymbol{v}(\cdot,\emptyset) = \delta^- \boldsymbol{v}(\cdot,\emptyset) = 0$. □

22 Proof of Theorem 20.4

22.1 In this section we present a proof of Theorem 20.4. This proof is in large parts constructive, i.e., we construct the Fréchet differential operator $D\boldsymbol{v}(tX,\cdot)$ for games in $pFNA_{\mathscr{T}}$ and we show that it satisfies the requirements of Theorem 20.4. The most interesting feature of the construction is the fact that the differential operator $D\boldsymbol{v}(tX,\cdot)$ is defined by the Dini derivative $\delta^+ \boldsymbol{v}(t,\cdot)$ (which for games in $pFNA_{\mathscr{T}}$ always exists as we shall show below). This fact will be used later in order to extend the diagonal value on $pFNA_{\mathscr{T}}$ to Aumann–Shapley values on spaces which strictly contain $pFNA_{\mathscr{T}}$ as does, for instance, the space of all F-differentiable games on $\mathscr{T}$. As in Sections 20 and 21, we assume here that $\mathscr{T}$ is a generated T_∞-tribe on X and we denote $\mathscr{A} = \mathscr{T}^\vee$. Our proof of Theorem 20.4 consists of a sequence of Lemmas.

22.2 We start with the following modified version of Lemma 24.2 from AUMANN & SHAPLEY [1974].

Lemma. *Let B be a Banach space and let B' be a closed subspace of B. If $q : B \longrightarrow [0,\infty[$ is a seminorm on B, if there exists a constant $\alpha \in \mathbb{R}$ such that for all $b \in B$ we have $q(b) \le \alpha \cdot \|b\|$ and if q vanishes on a spanning subset S of B', then q vanishes on the whole subspace B'.*

PROOF: If $b \in B'$, then for each $\varepsilon > 0$ there exists a $b_\varepsilon \in S$ such that $\|b - b_\varepsilon\| \le \varepsilon$. Since q is a seminorm on B and since it vanishes on S, we have

$$0 \le q(b) = q(b_\varepsilon + (b - b_\varepsilon)) \le q(b_\varepsilon) + q(b - b_\varepsilon) = q(b - b_\varepsilon) \le \alpha\|b - b_\varepsilon e\| \le \alpha \cdot \varepsilon.$$

Taking the limit $\varepsilon \to 0$, this implies $q(b) = 0$. □

22.3 The following result is fundamental when studying the games in $pFNA_{\mathcal{T}}$.

Lemma. *If $\boldsymbol{v} \in pFNA_{\mathcal{T}}$ and if $A \in \mathcal{T}$ then the functions $\delta^+\boldsymbol{v}(\cdot, A)$ and $\delta^-\boldsymbol{v}(\cdot, A)$ are Borel measurable on $[0,1]$, and for λ-almost all $t \in [0,1]$ we have*

$$\delta^+\boldsymbol{v}(t, A) = \frac{\partial h_{\boldsymbol{v},A}}{\partial \tau}(t, 0^+) = \delta^-\boldsymbol{v}(t, A). \tag{22.1}$$

PROOF: First note that $\delta^+\boldsymbol{v}(1, A) = 0 = \delta^-\boldsymbol{v}(1, A)$ because of (21.5) and (21.6). Since $h_{\boldsymbol{v},A}(1,\tau) = \boldsymbol{v}(X)$ for each $\tau \in [0,1]$ it follows that (22.1) holds for $t = 1$. Fix $t \in [0,1[$ and denote $I_t = [-t, 1-t[$. If $\boldsymbol{v} = \boldsymbol{m}^k$, $\boldsymbol{m}$ being a T_∞-measure and $k \in \mathbb{N}$, then, according to Theorem 6.2, we have

$$h_{\boldsymbol{v},A}(t,\tau) = \left[\int_X \min(1, t_\tau A(x))\, \mathrm{d}\check{\boldsymbol{m}}\,(x)\right]^k \tag{22.2}$$

whenever $t \in [0,1]$ and $\tau \in [0,1]$. Note that in this case, $h_{\boldsymbol{v},A}(t,\cdot)$ is continuously differentiable on I_t and satisfies

$$\frac{\partial h_{\boldsymbol{v},A}}{\partial \tau}(t, 0^+) = k \cdot \boldsymbol{m}^{k-1}(X) \cdot \boldsymbol{m}(A) \cdot t^{k-1}.$$

This shows that $\frac{\partial h_{\boldsymbol{v},A}}{\partial \tau}(\cdot, 0^+)$ is continuous and, subsequently, Borel measurable on $[0,1]$. Observe that the function $\boldsymbol{u} \longmapsto h_{\boldsymbol{u},A}(t,\tau)$ is linear. If $\boldsymbol{v} = p \circ \boldsymbol{m}$ where $\boldsymbol{m}$ is an n-vector T_∞-measure and p is a polynomial in n variables such that $p(0) = 0$, then, according to Lemma 17.3, it can be represented as a linear combination of positive integer powers of T_∞-measures on $\mathcal{T}$ and, therefore, $(t,\tau) \longmapsto h_{\boldsymbol{v},A}(t,\tau)$ is a linear combination of functions which are continuously differentiable with respect to τ on the interval I_t and whose partial derivatives $\frac{\partial h_{\boldsymbol{v},A}}{\partial \tau}$ at $(t, 0^+)$ are continuous with respect to t on $[0,1]$. Consequently, (22.1) is satisfied for each game in $pFNA_{\mathcal{T}}$

which is a polynomial in T_∞-measures on $\mathscr{T}$. This implies that for each game in $pFNA_{\mathscr{T}}$, which is a polynomial in T_∞-measures, the integrand in the definition of Δ_A (see (21.14)) vanishes. From Lemma 21.6 we know that Δ_A is a seminorm on $FBV_{\mathscr{T}}$. According to Lemma 21.6 and Lemma 22.2, Δ_A vanishes on $pFNA_{\mathscr{T}}$ because it vanishes on a spanning subset. Hence, for each $\boldsymbol{v} \in pFNA_{\mathscr{T}}$ and for each $A \in \mathscr{T}$ we have $\delta^+\boldsymbol{v}(t,A) = \delta^-\boldsymbol{v}(t,A)$ for λ-almost all $t \in [0,1]$. According to the definition of $\frac{\partial h_{\boldsymbol{v},A}}{\partial \tau}$, the equalities (22.1) are proven for all $\boldsymbol{v} \in pFNA_{\mathscr{T}}$. As shown above, for games in $pFNA_{\mathscr{T}}$ which are polynomials in T_∞-measures the derivative $\frac{\partial h_{\boldsymbol{v},A}}{\partial \tau}(t,0^+)$ is continuous (and therefore, Borel measurable), with respect to $t \in [0,1]$. If $\boldsymbol{v}$ is the limit (in $FBV_{\mathscr{T}}$) of a sequence of games $(\boldsymbol{v}_k)_{k\in\mathbb{N}} = (p_k \circ \boldsymbol{m}_k)_{k\in\mathbb{N}}$, where, for each $k \in \mathbb{N}$, $\boldsymbol{m}_k$ is an n_k-vector T_∞-measure and p_k is a polynomial in n_k variables such that $p_k(0) = 0$, then the sequence of functions $(h_{\boldsymbol{v}_k,A}(t,\tau))_{k\in\mathbb{N}}$ converges to $h_{\boldsymbol{v},A}(t,\tau)$ uniformly on $[0,1]^2$ because of (21.2). Also, from (22.1) and Lemma 21.5 we have

$$\begin{aligned}
\int_0^1 \left|\frac{\partial h_{\boldsymbol{v},A}}{\partial \tau}(t,0^+) - \frac{\partial h_{\boldsymbol{v}_k,A}}{\partial \tau}(t,0^+)\right| \mathrm{d}\lambda(t) &= \int_0^1 |\partial^+(\boldsymbol{v}-\boldsymbol{v}_k)(t,A)|\,\mathrm{d}\lambda(t) \\
&= \int_0^1 \partial^+(\boldsymbol{v}-\boldsymbol{v}_k)(t,A)\,\mathrm{d}\lambda(t) \\
&\le \|\boldsymbol{v}-\boldsymbol{v}_k\|,
\end{aligned}$$

showing that for λ-almost all $t \in [0,1]$

$$\lim_{k\to\infty} \frac{\partial h_{\boldsymbol{v}_k,A}}{\partial \tau}(t,0^+) = \frac{\partial h_{\boldsymbol{v},A}}{\partial \tau}(t,0^+).$$

As a consequence, $\frac{\partial h_{\boldsymbol{v},A}}{\partial \tau}(\cdot,0^+)$ is a λ-a.e. pointwise limit of a sequence of functions which are Borel measurable on $[0,1]$, and $\frac{\partial h_{\boldsymbol{v},A}}{\partial \tau}(\cdot,0^+)$ itself is Borel measurable on $[0,1]$. According to (22.1), this implies that $\delta^+\boldsymbol{v}(\cdot,A)$ and $\delta^-\boldsymbol{v}(\cdot,A)$ are Borel measurable on $[0,1]$. □

22.4 Definition. A game $\boldsymbol{v} \in FBV_{\mathscr{T}}$ satisfying

$$\delta^+\boldsymbol{v}(t,A) = \delta^-\boldsymbol{v}(t,A) \tag{22.3}$$

is called *Dini derivable at tX in the direction A*, and the common value of the upper and lower Dini derivatives of $\boldsymbol{v}$ at tX in the direction A is denoted $\delta\boldsymbol{v}(t,A)$ and is called *Dini derivative of $\boldsymbol{v}$ at tX in the direction A*. If $\boldsymbol{v}$ is Dini derivable at tX for λ-almost all $t \in [0,1]$ in each direction A, then $\boldsymbol{v}$ is called *Dini derivable*.

Lemma 22.3 implies that games in $pFNA_{\mathscr{T}}$ are Dini derivable. It will be shown later that the games in $pFNA_{\mathscr{T}}$ are not the only Dini derivable games, but that Dini derivability of all games in a symmetric subspace of $FBV_{\mathscr{T}}$ is not sufficient for the existence of a diagonal value on that subspace.

22.5 The following consequence of Lemma 22.3 will be of use.

Corollary. *For each $\boldsymbol{v} \in pFNA_{\mathscr{T}}$ and for each $A \in \mathscr{T}$ the function $\delta\boldsymbol{v}(\cdot, A)$ is λ-integrable and we have*

$$\int_0^1 |\delta\boldsymbol{v}(t, A)| \, \mathrm{d}\lambda(t) \leq \|\boldsymbol{v}\|. \tag{22.4}$$

PROOF: Since $\boldsymbol{v}$ is Dini derivable at tX for λ-almost all $t \in [0,1]$ in each direction A and the Dini derivative $\delta(\cdot, A)$ is Borel measurable on $[0,1]$ (cf. Lemma 22.3), the integral in (22.4) exists and we have $|\delta\boldsymbol{v}(\cdot, A)| = \partial^{+}\boldsymbol{v}(\cdot, A)$ λ-a.e. Hence, the inequality in (22.4) follows from Lemma 21.4. □

22.6 Now we are in the position to give a representation of the diagonal value φ on $pFNA_{\mathscr{T}}$ in terms of Dini derivatives. For this purpose denote

$$\Phi\boldsymbol{v}(A) = \int_0^1 \delta\boldsymbol{v}(t, A) \, \mathrm{d}\lambda(t) \tag{22.5}$$

whenever $\boldsymbol{v} \in pFNA_{\mathscr{T}}$ and $A \in \mathscr{T}$. Note that, according to Lemma 22.3 and Corollary 22.5, $\Phi\boldsymbol{v}(A)$ is well-defined and

$$|\Phi\boldsymbol{v}(A)| \leq \|\boldsymbol{v}\| \tag{22.6}$$

Lemma. *For each $\boldsymbol{v} \in pFNA_{\mathscr{T}}$ and for each $A \in \mathscr{T}$ we have*

$$\varphi\boldsymbol{v}(A) = \Phi\boldsymbol{v}(A). \tag{22.7}$$

PROOF: First, according to Lemma 22.3, we have

$$\Phi\boldsymbol{v}(A) = \int_0^1 \delta\boldsymbol{v}(t, A) \, \mathrm{d}\lambda(t) = \int_0^1 \frac{\partial h_{\boldsymbol{v},A}}{\partial \tau}(t, 0^{+}) \, \mathrm{d}\lambda(t). \tag{22.8}$$

If $\boldsymbol{v}$ is a positive integer power of a T_{∞}-measure on $\mathscr{T}$, say $\boldsymbol{v} = \boldsymbol{m}^k$, then (22.2) is satisfied and, taking into account the diagonal formula (18.6), we get

$$\Phi\boldsymbol{v}(A) = \boldsymbol{m}^{k-1}(X) \cdot \boldsymbol{m}(A) = \varphi\boldsymbol{v}(A).$$

If $\boldsymbol{v} = p \circ \boldsymbol{m}$, where $\boldsymbol{m}$ is an n-vector T_{∞}-measure and p is a polynomial in n variables such that $p(0) = 0$, then, by Lemma 17.3, $\boldsymbol{v}$ is a linear combination of positive integer powers of T_{∞}-measures and, using the linearity of $h_{\boldsymbol{v},A}$ with respect to $\boldsymbol{v}$ and formula (18.6), we obtain that (22.7) holds in this case too. Note that $\delta\boldsymbol{v}(t, A)$ is linear with respect to $\boldsymbol{v}$ in the sense that

$$\delta(\boldsymbol{v}_1 + \boldsymbol{v}_2)(t, A) = \delta\boldsymbol{v}_1(t, A) + \delta\boldsymbol{v}_2(t, A),$$

whenever both Dini derivatives $\delta \boldsymbol{v}_1(t, A)$ and $\delta \boldsymbol{v}_2(t, A)$ exist. By (22.6), the function $\boldsymbol{v} \longmapsto \Phi\boldsymbol{v}(A)$ is linear and bounded on $pFNA_{\mathcal{T}}$, i.e., it is continuous. Since for each $A \in \mathcal{T}$, $\boldsymbol{v} \longmapsto \varphi\boldsymbol{v}(A)$ and $\boldsymbol{v} \longmapsto \Phi\boldsymbol{v}(A)$ are continuous linear functionals on $pFNA_{\mathcal{T}}$ which coincide on a spanning subset of $pFNA_{\mathcal{T}}$, it follows that (22.7) is satisfied for each $\boldsymbol{v} \in pFNA_{\mathcal{T}}$. □

22.7 The diagonal values on $pFNA_{\mathcal{T}}$ are T_∞-additive by definition. Therefore, Lemma 22.6 implies that for each pair (A, B) of T_∞-disjoint fuzzy coalitions we have

$$\delta \boldsymbol{v}(\cdot, A\,\mathbf{S}_\infty\, B) = \delta \boldsymbol{v}(\cdot, A) + \delta \boldsymbol{v}(\cdot, B) \qquad \lambda\text{-a.e.} \tag{22.9}$$

However, we can prove a stronger result as follows:

Lemma. *If $\boldsymbol{v} \in pFNA_{\mathcal{T}}$, then for λ-almost all $t \in [0,1]$ the functions $\delta\boldsymbol{v}(t,\cdot)$ and $\varphi\boldsymbol{v}$ are T_∞-measures on $\mathcal{T}$.*

PROOF: Let $\mathcal{S}$ be the family of nondecreasing sequences of fuzzy sets in $\mathcal{T}$. Define the function $\Lambda : pFNA_{\mathcal{T}} \longrightarrow [0,\infty]$ by

$$\Lambda \boldsymbol{v} = \int_0^1 \sup_{(A_n)\in\mathcal{S}} [\limsup_{n\to\infty} \delta\boldsymbol{v}(t, A_n) - \liminf_{n\to\infty} \delta\boldsymbol{v}(t, A_n)]\, \mathrm{d}\lambda(t).$$

where the supremum is taken over all sequences $(A_n)_{n\in\mathbb{N}} \in \mathcal{S}$. It is easily seen that Λ is a seminorm of $pFNA_{\mathcal{T}}$. By Lemma 22.3, Λ vanishes on the set of linear combinations of positive integer powers of T_∞-measures. Note that, according to (21.12), for each $t \in [0,1]$ and for each sequence $(A_n)_{n\in\mathbb{N}} \in \mathcal{S}$ we have

$$\max\left[|\liminf_{n\to\infty} \delta\boldsymbol{v}(t, A_n)|, |\limsup_{n\to\infty} \delta\boldsymbol{v}(t, A_n)|\right] \le \Delta^+\boldsymbol{v}(t),$$

and thus

$$\Lambda\boldsymbol{v} \le \int_0^1 \sup_{(A_n)\in\mathcal{S}} |\liminf_{n\to\infty} \delta\boldsymbol{v}(t, A_n)|\, \mathrm{d}\lambda(t) + \int_0^1 \sup_{(A_n)\in\mathcal{S}} |\limsup_{n\to\infty} \delta\boldsymbol{v}(t, A_n)|\, \mathrm{d}\lambda(t).$$

Hence, for each $\boldsymbol{v} \in pFNA_{\mathcal{T}}$

$$\Lambda\boldsymbol{v} \le 2 \cdot \int_0^1 \Delta^+\boldsymbol{v}(t)\, \mathrm{d}\lambda(t) \le 2 \cdot \|\boldsymbol{v}\|,$$

because of Lemma 21.5. Combining this inequality and Lemma 22.2, it follows that Λ vanishes on $pFNA_{\mathcal{T}}$ since it vanishes on a spanning subset of $pFNA_{\mathcal{T}}$. Therefore, there is a set $M \subseteq [0,1]$ such that $\lambda(M) = 1$ and such that for each $t \in M$ and for each $(A_n)_{n\in\mathbb{N}} \in \mathcal{S}$ the sequence $(\delta\boldsymbol{v}(t, A_n))_{n\in\mathbb{N}}$ converges. Define the function $\Lambda_1 : pFNA_{\mathcal{T}} \longrightarrow [0,\infty]$ by

$$\Lambda_1\boldsymbol{v} = \int_0^1 \sup_{(A_n)\in\mathcal{S}} \left[|\lim_{n\to\infty} \delta\boldsymbol{v}(t, A_n) - \delta\boldsymbol{v}(t, \mathop{\mathbf{S}_0}_{n=1}^{\infty} A_n)|\right] \mathrm{d}\lambda(t).$$

Since the limit in this formula exists λ-a.e. the function Λ_1 is well-defined and it is a seminorm on $pFNA_{\mathscr{T}}$. Using Lemma 22.3, again it is easily seen that $\delta v(t,\cdot)$ is a T_∞-measure whenever v is a linear combination of positive integer powers of T_∞-measures on $\mathscr{T}$. Therefore, the integrand in the definition of Λ_1 vanishes on the spanning subset of $pFNA_{\mathscr{T}}$ consisting of all linear combinations of positive integer powers of T_∞-measures on $\mathscr{T}$. By Lemma 21.4 we obtain $\Lambda_1 v \leq 2 \cdot \|v\|$ in the same way as before for Λ. Therefore, Lemma 22.2 can be applied to Λ_1 showing that Λ_1 vanishes on $pFNA_{\mathscr{T}}$. Consequently, there exists a subset $M^* \subseteq [0,1]$ such that $\lambda(M^*) = 1$ and such that for each sequence $(A_n)_{n\in\mathbb{N}} \in \mathscr{S}$ and for each $t \in M^*$ we have

$$\lim_{n\to\infty} \delta v(t, A_n) = \delta v(t, \mathbf{S}_0{}_{n=1}^{\infty} A_n).$$

This implies that for λ-almost all $t \in [0,1]$ the function $\delta v(t,\cdot)$ is continuous from the left on $\mathscr{T}$. Define the function $\Lambda_2 : pFNA_{\mathscr{T}} \longrightarrow [0,1]$ by

$$\Lambda_2 v = \bar{\int_0^1} \sup\,[|\delta v(t,A) + \delta v(t,B) - \delta v(t, A\,\mathbf{S}_\infty\, B)|]\ \mathrm{d}\lambda(t),$$

where the supremum is taken over all pairs (A, B) of T_∞-disjoint fuzzy sets in $\mathscr{T}$. Similar reasoning to before for Λ_1 shows that Λ_2 vanishes on $pFNA_{\mathscr{T}}$ and, consequently, that for λ-almost all $t \in [0,1]$ the function $\delta v(t,\cdot)$ is finitely T_∞-additive on $\mathscr{T}$. Since continuity from the left together with finite T_∞-additivity implies countable T_∞-additivity (cf. Proposition 9.1), $\delta v(t,\cdot)$ is a T_∞-measure for λ-almost all $t \in [0,1]$. According to Lemma 22.6, φv is also a T_∞-measure. □

22.8 For each $v \in pFNA_{\mathscr{T}}$ we denote by $\mathscr{J}(v)$ the set of all $t \in [0,1]$ such that $\delta v(t,\cdot)$ is a T_∞-measure. Lemma 22.7 combined with Corollary 10.4 gives for each $t \in \mathscr{J}(v)$ and each $A \in \mathscr{T}$

$$\delta v(t, A) = \int_X A(x)\,\mathrm{d}[\delta v(t,\cdot)]^\vee(x) \tag{22.10}$$

where $[\delta v(t,\cdot)]^\vee$, as usual, denotes the restriction of $\delta v(t,\cdot)$ to $\mathscr{A}$ (which is a σ-additive measure on $\mathscr{A}$). This suggests a way of extending $\delta v(t,\cdot)$ to the whole space $\mathscr{M}$. Define $Dv(tX,\cdot) : \mathscr{M} \longrightarrow \mathbb{R}$ by

$$Dv(tX, f) = \int_X f(x)\,\mathrm{d}[\delta v(t,\cdot)]^\vee(x). \tag{22.11}$$

This function is well-defined whenever $t \in \mathscr{J}(v)$. For each $t \in \mathscr{J}(v)$ the function $Dv(tX,\cdot)$ is linear on $\mathscr{M}$ and obviously satisfies:

$$|Dv(tX, f)| \leq \|\delta v(t,\cdot)\| \cdot \|f\|. \tag{22.12}$$

Hence, whenever it exists, $Dv(tX,\cdot)$ is a continuous linear operator on $\mathcal{M}$. The following result shows that all games in $pFNA_{\mathcal{T}}$ are continuously F-differentiable.

Lemma. *If $v \in pFNA_{\mathcal{T}}$, then for λ-almost all $t \in [0,1]$ the operator $Dv(tX,\cdot)$ equals the Fréchet differential of v at tX, i.e., (20.2) holds λ-a.e.*

PROOF: Suppose $t \in \mathcal{J}(v)$. Then

$$\begin{aligned} 0 &\leq \limsup_{\|f\|\searrow 0} \frac{|v(tX+f)-v(tX)-Dv(tX,f)|}{\|f\|} \\ &\leq \limsup_{\|f\|\searrow 0} \frac{|v(tX+f)-v(tX)|}{\|f\|} + \limsup_{\|f\|\searrow 0} \frac{|Dv(tX,f)|}{\|f\|} \\ &= \Delta^+ v + \limsup_{\|f\|\searrow 0} \left| Dv\left(tX, \frac{f}{\|f\|}\right)\right|. \end{aligned}$$

For $f \in \mathcal{M}$ define the two functions

$$f^+ = \frac{|f|+f}{2}, \qquad f^- = \frac{|f|-f}{2}. \tag{22.13}$$

Obviously, $f = f^+ - f^-$, $\|f\| \geq \max(\|f^+\|, \|f^-\|)$ and, whenever $f \neq 0$, $\frac{f^+}{\|f\|}$ as well as $\frac{f^-}{\|f\|}$ belong to $\mathcal{T}$. By (22.10), (22.11) and Lemma 22.3 we obtain for each $t \in \mathcal{J}(v)$

$$\begin{aligned} \left|Dv\left(tX, \frac{f^+}{\|f\|}\right)\right| &= \left|Dv\left(tX, \frac{f^+}{\|f\|}\right) - Dv\left(tX, \frac{f^-}{\|f\|}\right)\right| \\ &\leq \left|Dv\left(tX, \frac{f^+}{\|f\|}\right)\right| + \left|Dv\left(tX, \frac{f^-}{\|f\|}\right)\right| \\ &= \left|\delta v\left(t, \frac{f^+}{\|f\|}\right)\right| + \left|\delta v\left(t, \frac{f^-}{\|f\|}\right)\right| \\ &\leq \left|\partial^+ v\left(t, \frac{f^+}{\|f\|}\right)\right| + \left|\partial^+ v\left(t, \frac{f^-}{\|f\|}\right)\right| \\ &\leq 2\cdot\Delta^+ v(t), \end{aligned}$$

and thus Lemma 21.4 implies that for all $f \in \mathcal{M}$ and for λ-almost all $t \in [0,1]$

$$\left|Dv\left(tX, \frac{f^+}{\|f\|}\right)\right| \leq 2\cdot\Delta^+ v(t).$$

Hence, for λ-almost all $t \in [0,1]$

$$0 \leq \limsup_{\|f\|\searrow 0} \frac{|v(tX+f)-v(tX)-Dv(tX,f)|}{\|f\|} \leq 3\cdot\Delta^+ v(t). \tag{22.14}$$

Define the function $\Psi : pFNA_{\mathscr{T}} \longrightarrow [0,\infty]$, $\boldsymbol{v} \longmapsto \Psi\boldsymbol{v}$, by

$$\Psi\boldsymbol{v} = \int_0^1 \limsup_{\|f\|\searrow 0} \frac{|\boldsymbol{v}(tX+f) - \boldsymbol{v}(tX) - D\boldsymbol{v}(tX,f)|}{\|f\|} \, \mathrm{d}\lambda(t).$$

Since $D\boldsymbol{v}(tX,f)$ is linear with respect to $\boldsymbol{v}$, it follows that Ψ is a seminorm on $pFNA_{\mathscr{T}}$. By (22.14) we have

$$\Psi\boldsymbol{v} \leq 3 \cdot \|\boldsymbol{v}\|, \tag{22.15}$$

for each $\boldsymbol{v} \in pFNA_{\mathscr{T}}$. If $\boldsymbol{v}$ is a linear combination of positive integer powers of T_∞-measures on $\mathscr{T}$, then

$$\limsup_{\|f\|\searrow 0} \frac{|\boldsymbol{v}(tX+f) - \boldsymbol{v}(tX) - D\boldsymbol{v}(tX,f)|}{\|f\|} = 0,$$

because of Corollary 10.4. Lemma 17.3 implies that Ψ vanishes on a spanning subset of $pFNA_{\mathscr{T}}$. Taking into account (22.15) and Lemma 22.2, Ψ vanishes on $pFNA_{\mathscr{T}}$. As a consequence,

$$\begin{aligned} 0 &\leq \liminf_{\|f\|\searrow 0} \frac{|\boldsymbol{v}(tX+f) - \boldsymbol{v}(tX) - D\boldsymbol{v}(tX,f)|}{\|f\|} \\ &\leq \limsup_{\|f\|\searrow 0} \frac{|\boldsymbol{v}(tX+f) - \boldsymbol{v}(tX) - D\boldsymbol{v}(tX,f)|}{\|f\|} \\ &= 0 \end{aligned}$$

which means that

$$\lim_{\|f\|\searrow 0} \frac{|\boldsymbol{v}(tX+f) - \boldsymbol{v}(tX) - D\boldsymbol{v}(tX,f)|}{\|f\|} = 0$$

for λ-almost all $t \in [0,1]$. □

22.9 Lemma 22.8 completes the proof of Theorem 20.4 as follows: Games in $pFNA_{\mathscr{T}}$ are F-differentiable and, by (22.12), the F-differential operator at tX is continuous on $\mathscr{M}$ for λ-almost all $t \in [0,1]$. If $A \in \mathscr{T}$, then (22.11) and (22.10) imply that $D\boldsymbol{v}(tX,A) = \delta\boldsymbol{v}(t,A)$ whenever both the Fréchet differential $D\boldsymbol{v}(tX,A)$ and the Dini derivative $\delta\boldsymbol{v}(t,A)$ exist. Hence, for each $A \in \mathscr{T}$ the function $t \longmapsto D\boldsymbol{v}(tX,A)$ is λ-integrable (cf. Lemma 22.3). Therefore, each game $\boldsymbol{v} \in pFNA_{\mathscr{T}}$ is continuously differentiable on the diagonal. Combining Lemma 22.6 and (22.11), formula (20.3) holds, and Theorem 20.4 is completely proved. □

23 Diagonal Value for Gâteaux Differentiable Games

23.1 In this section we show that if $\mathcal{T}$ is a generated T_∞-tribe, then the diagonal value φ on $pFNA_{\mathcal{T}}$ can be extended to a subspace of $FBV_{\mathcal{T}}$ which contains all F-differentiable games. We also show that the diagonal value on $pFNA_{\mathcal{T}}$ can be extended to a continuous *semivalue* on the space of all *Gâteaux differentiable* games on $\mathcal{T}$. Recall that a semivalue is an operator from a symmetric subspace of $FBV_{\mathcal{T}}$ to $FBA_{\mathcal{T}}$ which is linear, positive and symmetric, but not necessarily efficient.

Definition. A *game* $\boldsymbol{v} \in FBV_{\mathcal{T}}$ is called *Gâteaux differentiable on the diagonal* (or briefly G-*differentiable*) if for λ-almost all $t \in [0,1]$:

(i) $\boldsymbol{v}$ is Dini derivable at tX in each direction $A \in \mathcal{T}$;

(ii) the function $\delta\boldsymbol{v}(t,\cdot)$ is a finitely T_∞-additive function on $\mathcal{T}$.

Note that for $\boldsymbol{v} \in FBV_{\mathcal{T}}$, the function $g_{\boldsymbol{v}}$ is differentiable λ-a.e. and that its derivative $g'_{\boldsymbol{v}}$ is Borel measurable. If we have

$$\int_0^1 g'_{\boldsymbol{v}}(t)\,\mathrm{d}\lambda(t) = \boldsymbol{v}(X) \tag{23.1}$$

then $\boldsymbol{v}$ will be called *diagonally efficient*. $GDIFF_{\mathcal{T}}$ and $GDIFFE_{\mathcal{T}}$ will denote the sets of games in $FBV_{\mathcal{T}}$ which are Gâteaux differentiable and diagonally efficient Gâteaux differentiable, respectively.

23.2 The following result shows that G-differentiability, F-differentiability and continuous F-differentiability are equivalent. In view of Theorem 20.4, it also shows that $pFNA_{\mathcal{T}} \subseteq GDIFFE_{\mathcal{T}}$.

Proposition. *For $\boldsymbol{v} \in FBV_{\mathcal{T}}$ the following are equivalent:*

(i) $\boldsymbol{v}$ is G-differentiable;

(ii) $\boldsymbol{v}$ is F-differentiable;

(iii) $\boldsymbol{v}$ is continuously F-differentiable.

Moreover, if $\boldsymbol{v}$ is G-differentiable, then the Fréchet differential operator $D\boldsymbol{v}(tX,\cdot)$ exists whenever $\delta\boldsymbol{v}(t,\cdot)$ is finite, in which case we have for each $A \in \mathcal{T}$

$$D\boldsymbol{v}(tX,A) = \delta\boldsymbol{v}(t,A). \tag{23.2}$$

PROOF: First, we show that if $\boldsymbol{v}$ is G-differentiable, then for each $A \in \mathscr{T}$ the function $\delta\boldsymbol{v}(\cdot, A)$ is Borel measurable. If $\delta\boldsymbol{v}(\cdot, A)$ is Borel measurable, then it is λ-integrable on $[0,1]$ by Lemma 21.5. In order to show the Borel measurability, note that by Lemma 21.8 the function $\delta\boldsymbol{v}(\cdot, A)$ is λ-a.e. finite. For each $t \in]0,1]$ define the function $f_t : [0,1] \longrightarrow \mathbb{R}$ by

$$f_t(\tau) = \boldsymbol{v}(tX\,\mathbf{S}_\infty\,\tau A).$$

Since for each $t \in]0,1]$

$$f_t(\tau) = \boldsymbol{v}(tX) + \frac{\boldsymbol{v}(tX\,\mathbf{S}_\infty\,\tau A) - \boldsymbol{v}(tX)}{\tau} \cdot \tau,$$

it follows that whenever $\delta\boldsymbol{v}(t, A)$ is finite, then we have

$$\lim_{\tau \searrow 0} f_t(\tau) = \boldsymbol{v}(tX) = f_t(0),$$

i.e., f_t is continuous from the right at $t = 0$. Note that

$$\delta\boldsymbol{v}(t, A) = \lim_{\tau \searrow 0} \frac{f_t(\tau) - f_t(0)}{\tau}$$

and that for a fixed $\tau \in [0,1]$, $f_t(\tau) - f_t(0)$ is Borel measurable as a function of t. Whenever $\delta\boldsymbol{v}(t, A)$ is finite, it is sufficient to take the limit in the latter equality for a sequence of rational numbers $(\tau_n)_{n\in\mathbb{N}}$ converging decreasingly to zero because f_t is continuous from the right at $\tau = 0$. This shows that $\delta\boldsymbol{v}(\cdot, A)$ is a λ-a.e. pointwise limit of a sequence of Borel measurable functions and thus $\delta\boldsymbol{v}(\cdot, A)$ is itself Borel measurable. We are now ready to prove (i) $\Rightarrow$ (iii). Assume that $\boldsymbol{v}$ is G-differentiable. It is sufficient to show that whenever $\delta\boldsymbol{v}(t, \cdot)$ exists and is finite, the function $\delta\boldsymbol{v}(t, \cdot)$ can be extended to a continuous linear operator $D\boldsymbol{v}(tX, \cdot) : \mathscr{M} \longrightarrow \mathbb{R}$ satisfying condition (23.2). To this end, note that whenever $m, n \in \mathbb{N}$, $m \leq n$ and $A \in \mathscr{T}$, we have

$$\delta\boldsymbol{v}\left(t, \frac{m}{n} \cdot A\right) = m \cdot \delta\boldsymbol{v}\left(t, \frac{1}{n} \cdot A\right)$$

since $\delta\boldsymbol{v}(t, \cdot)$ is finitely T_∞-additive and the family of fuzzy sets $(A_1, A_2, \ldots, A_n)$, where each A_i is given by $A_i = \frac{1}{n} \cdot A$, is T_∞-disjoint. Observe that for $m = n$ this gives

$$\delta\boldsymbol{v}(t, A) = n \cdot \delta\boldsymbol{v}(t, \frac{1}{n} \cdot A),$$

implicitly showing that for each rational number $\alpha \in [0,1]$ we have

$$\delta\boldsymbol{v}(t, \alpha A) = \alpha \cdot \delta\boldsymbol{v}(t, A).$$

If $\beta \in [0,1]$ then there exists a sequence $(\alpha_k)_{k\in\mathbb{N}}$ of rational numbers in $[0,1]$ such that $(\alpha_k)_{k\in\mathbb{N}} \nearrow \beta$. For each $k \in \mathbb{N}$

$$\delta\boldsymbol{v}(t,\beta A) = \delta\boldsymbol{v}(t,(\beta-\alpha_k)\cdot A) + \alpha_k \cdot \delta\boldsymbol{v}(t,A)$$

and since $((\beta-\alpha_k)\cdot A)_{k\in\mathbb{N}}$ converges uniformly to zero,

$$\lim_{k\to\infty} \delta\boldsymbol{v}(t,(\beta-\alpha_k)\cdot A) = 0$$

by Lemma 21.5. Therefore, using the fact that $\alpha_k \in \mathbb{Q}\cap[0,1]$ we get

$$\delta\boldsymbol{v}(t,\beta A) = \lim_{k\to\infty} \alpha_k \cdot \delta\boldsymbol{v}(t,A) = \beta\cdot\delta\boldsymbol{v}(t,A),$$

proving that $\delta\boldsymbol{v}(t,\cdot)$ is positively homogeneous. If $f \in \mathscr{M}$, and $f \neq 0$, then denote

$$D\boldsymbol{v}(tX,f) = \|f\|\cdot\left[\delta\boldsymbol{v}\left(t,\frac{f^+}{\|f\|}\right) - \delta\boldsymbol{v}\left(t,\frac{f^-}{\|f\|}\right)\right]$$

where f^+ and f^- are given by (22.13). Then $(-f)^+ = f^-$ and $(-f)^- = f^+$ imply that $D\boldsymbol{v}(tX,-f) = -D\boldsymbol{v}(tX,f)$ showing that $D\boldsymbol{v}(tX,\cdot)$ is homogeneous. If $f,g \in \mathscr{M}$ are nonnegative functions, then

$$\begin{aligned}
D\boldsymbol{v}(tX,f+g) &= \|f+g\|\cdot\delta\boldsymbol{v}\left(t,\frac{f+g}{\|f+g\|}\right)\\
&= \|f+g\|\cdot\left[\delta\boldsymbol{v}\left(t,\frac{f}{\|f+g\|}\right) + \delta\boldsymbol{v}\left(t,\frac{g}{\|f+g\|}\right)\right]\\
&= D\boldsymbol{v}(tX,f) + D\boldsymbol{v}(tX,g)
\end{aligned}$$

since $D\boldsymbol{v}(tX,\cdot)$ is homogeneous and $\delta\boldsymbol{v}(t,\cdot)$ is finitely T_∞-additive. Now, if $f,g \in \mathscr{M}$ are arbitrary functions, observe that

$$(f+g)^+ + f^- + g^- = (f+g)^- + f^+ + g^+,$$

where all functions involved are nonnegative. Hence,

$$\begin{aligned}
D\boldsymbol{v}(tX,(f+g)^+) - D\boldsymbol{v}(tX,(f+g)^-) &= D\boldsymbol{v}(tX,f^+) - D\boldsymbol{v}(tX,f^-)\\
&\quad + D\boldsymbol{v}(tX,g^+) - D\boldsymbol{v}(tX,g^-)\\
&= D\boldsymbol{v}(tX,f) + D\boldsymbol{v}(tX,g)
\end{aligned}$$

showing that $D\boldsymbol{v}(tX,\cdot)$ is additive and, therefore, linear. Since by Lemma 21.5 there exists a constant $K_t > 0$ such that for each $A \in \mathscr{T}$, $|D\boldsymbol{v}(tX,A)| \leq K_t\cdot\|A\|$, it follows that for each $f \in \mathscr{M}$

$$\begin{aligned} |D\boldsymbol{v}(tX,f) - D\boldsymbol{v}(tX,\emptyset)| &= \|f\| \cdot \left| D\boldsymbol{v}\left(tX, \frac{f^+}{\|f\|}\right) - D\boldsymbol{v}\left(tX, \frac{f^-}{\|f\|}\right) \right| \\ &\leq \|f\| \cdot \left[\left| D\boldsymbol{v}\left(tX, \frac{f^+}{\|f\|}\right) \right| + \left| D\boldsymbol{v}\left(tX, \frac{f^-}{\|f\|}\right) \right| \right] \\ &\leq K_t \cdot \|f\| \cdot \left[\frac{\|f^+\|}{\|f\|} + \frac{\|f^-\|}{\|f\|} \right] \\ &\leq 2K_t \cdot \|f\|, \end{aligned}$$

because of $\frac{f^+}{\|f\|}, \frac{f^-}{\|f\|} \in \mathscr{T}$. Hence, $D\boldsymbol{v}(tX,\cdot)$ is continuous at $\emptyset$ and, consequently, continuous on $\mathscr{M}$. It is clear that $D\boldsymbol{v}(tX,\cdot)$ defined above satisfies (23.2) and, therefore, the game $\boldsymbol{v}$ is continuously F-differentiable. Obviously, we have (iii) $\Rightarrow$ (ii). It remains to show that (ii) $\Rightarrow$ (i). Assume that $\boldsymbol{v}$ is F-differentiable and that $D\boldsymbol{v}(tX,\cdot)$ exists for some $t \in [0,1]$. Then for each $A \in \mathscr{T}$

$$\begin{aligned} 0 &= \liminf_{\|f\| \searrow 0} \frac{|\boldsymbol{v}(tX+f) - \boldsymbol{v}(tX) - D\boldsymbol{v}(tX,f)|}{\|f\|} \\ &\leq \liminf_{\tau \searrow 0} \frac{|\boldsymbol{v}(tX+\tau A) - \boldsymbol{v}(tX) - D\boldsymbol{v}(tX,\tau A)|}{\tau} \\ &\leq \limsup_{\tau \searrow 0} \frac{|\boldsymbol{v}(tX+\tau A) - \boldsymbol{v}(tX) - D\boldsymbol{v}(tX,\tau A)|}{\tau} \\ &\leq \limsup_{\|f\| \searrow 0} \frac{|\boldsymbol{v}(tX+f) - \boldsymbol{v}(tX) - D\boldsymbol{v}(tX,f)|}{\|f\|} \\ &= 0. \end{aligned}$$

Therefore, the Dini derivative $\delta\boldsymbol{v}(t,A)$ in each direction A exists, (23.2) is satisfied and $\delta\boldsymbol{v}(t,\cdot)$ is T_∞-additive. Obviously, $\delta\boldsymbol{v}(t,\cdot)$ is finitely T_∞-additive since $D\boldsymbol{v}(tX,\cdot)$ is linear. □

23.3 It was noted above that $pFNA_{\mathscr{T}} \subseteq GDIFF_{\mathscr{T}}$. In fact, we have the following even stronger result:

Corollary. $pFNA_{\mathscr{T}} \subseteq GDIFFE_{\mathscr{T}}$.

PROOF: It is sufficient to show that each $\boldsymbol{v} \in pFNA_{\mathscr{T}}$ is diagonally efficient. Let φ be the diagonal value on $pFNA_{\mathscr{T}}$. In view of Theorem 20.3 and of Proposition 23.2 we have

$$\int_0^1 g_{\boldsymbol{v}}'(t)\,\mathrm{d}\lambda(t) = \int_0^1 \delta\boldsymbol{v}(t,X)\,\mathrm{d}\lambda(t) = \int_0^1 D\boldsymbol{v}(tX,X)\,\mathrm{d}\lambda(t) = \varphi\boldsymbol{v}(X) = \boldsymbol{v}(X).$$

□

23.4 A game in $FBV_{\mathscr{T}}$ may be Dini derivable without being G-differentiable.

Example. Define the game $\boldsymbol{v} : \mathscr{T} \longrightarrow \mathbb{R}$ by $\boldsymbol{v}(A) = \|A\|$. This game is monotone; hence, it belongs to $FBV_{\mathscr{T}}$. It is also Dini derivable since $\delta\boldsymbol{v}(t, A) = \boldsymbol{v}(A)$ for all $t \in]0, 1[$ and $A \in \mathscr{T}$. However, $\boldsymbol{v}$ is not finitely T_∞-additive and, therefore, it is not G-differentiable.

23.5 Games in $pFNA_{\mathscr{T}}$ are F-differentiable and, hence, G-differentiable and diagonally efficient. However, there are games in $GDIFF_{\mathscr{T}}$ which are not diagonally efficient even if they are F-differentiable and, hence, G-differentiable.

Example. Put $X = [0,1]$ and let $\mathscr{T}$ be the T_∞-tribe of all Borel measurable fuzzy subsets of X. Define the game $\boldsymbol{v} : \mathscr{T} \longrightarrow \mathbb{R}$ by $\boldsymbol{v} = f \circ \hat{\lambda}$, where $f : [0,1] \longrightarrow \mathbb{R}$ is of bounded variation such that $f(0) = 0$, and $\hat{\lambda}$ is the T_∞-measure on $\mathscr{T}$ defined by

$$\hat{\lambda}(A) = \int_0^1 A(x)\, \mathrm{d}\lambda(x). \tag{23.3}$$

If $A \in \mathscr{T}$ and $\hat{\lambda}(A) = 0$, then $\boldsymbol{v}$ is Dini derivable in the direction A since $\delta\boldsymbol{v}(t, A) = 0$ for all $t \in [0, 1]$. If $\hat{\lambda}(A) \neq 0$ and $t \in]0, 1[$, then we have

$$\frac{\boldsymbol{v}(tX + \tau A) - \boldsymbol{v}(tX)}{\tau} = \hat{\lambda}(A) \cdot \frac{f(t + \tau\hat{\lambda}(A)) - f(t)}{\tau \cdot \hat{\lambda}(A)},$$

for each $\tau \in]0, 1 - t]$. Therefore, if f is differentiable at t, then

$$\delta\boldsymbol{v}(t, A) = f'(t) \cdot \hat{\lambda}(A),$$

and this shows that $\boldsymbol{v} \in GDIFF_{\mathscr{T}}$. It is easily seen that $\boldsymbol{v}$ is G-differentiable and that the corresponding differential operator is exactly

$$D\boldsymbol{v}(tX, g) = f'(t) \cdot \int_0^1 g(x)\, \mathrm{d}\lambda(x), \tag{23.4}$$

whenever f is differentiable at t and $g \in \mathscr{M}$. By Theorem 17.2, *if f is absolutely continuous on* $[0, 1]$, then $\boldsymbol{v} \in pFNA_{\mathscr{T}}$ and, therefore, $\boldsymbol{v} \in GDIFFE_{\mathscr{T}}$. However, *if f is not absolutely continuous* or if f does not satisfy sufficient *smoothness* conditions, *then $\boldsymbol{v}$ may not belong to* $GDIFFE_{\mathscr{T}}$. For instance, consider the function $f : [0, 1] \longrightarrow [0, 1]$ given by

$$f(t) = \begin{cases} 0 & \text{if } \ x \in [0, 0.5[\,, \\ 1 & \text{if } \ x \in [0.5, 1]. \end{cases} \tag{23.5}$$

Then $g_{\boldsymbol{v}} = f$ and

$$\int_0^1 g'_{\boldsymbol{v}}(t)\, \mathrm{d}\lambda(t) = 0 \neq 1 = f(1) = \boldsymbol{v}(X),$$

showing that $\boldsymbol{v}$ does not necessarily belong to $GDIFFE_{\mathcal{T}}$ even if $\boldsymbol{v}$ is F-differentiable.

23.6 The following result shows that the diagonal value on $pFNA_{\mathcal{T}}$ can be extended to a continuous *diagonal* semivalue on $GDIFF_{\mathcal{T}}$, and that $GDIFFE_{\mathcal{T}}$ is the largest subspace of $GDIFF_{\mathcal{T}}$ on which this semivalue is indeed an Aumann–Shapley value.

In what follows, differentiability of a function $f : [0,1] \longrightarrow \mathbb{R}$ at the points 0 and 1 will be used as a shortcut for the existence of finite right and left derivatives $f'(0^+)$ and $f'(1^-)$, respectively.

Theorem. *Let $\mathcal{T}$ be a generated T_∞-tribe on X. Then we have:*

(i) *$GDIFF_{\mathcal{T}}$ is a closed symmetric subspace of $FBV_{\mathcal{T}}$;*

(ii) *$GDIFFE_{\mathcal{T}}$ is a closed symmetric (proper) subspace of $GDIFF_{\mathcal{T}}$;*

(iii) *$GDIFF_{\mathcal{T}}$ contains the closed subspace of $FBV_{\mathcal{T}}$ spanned by the games $\boldsymbol{v} = f \circ \boldsymbol{m}$, where $\boldsymbol{m}$ is a probability T_∞-measure on $\mathcal{T}$ and $f : [0,1] \longrightarrow \mathbb{R}$ is a function of bounded variation with $f(0) = 0$;*

(iv) *$GDIFFE_{\mathcal{T}}$ contains the closed subspace of $FBV_{\mathcal{T}}$ spanned by the games $\boldsymbol{v} = f \circ \boldsymbol{m}$, where $\boldsymbol{m}$ is a probability T_∞-measure on $\mathcal{T}$, and $f : [0,1] \longrightarrow \mathbb{R}$ is a function of bounded variation which is differentiable at each point $t \in [0,1]$ and satisfies $f(0) = 0$;*

(v) *the function $\theta : GDIFF_{\mathcal{T}} \longrightarrow pFBA_{\mathcal{T}}$, $\boldsymbol{v} \longmapsto \theta\boldsymbol{v}$ given by*

$$\theta\boldsymbol{v}(A) = \int_0^1 \delta\boldsymbol{v}(t, A)\, \mathrm{d}\lambda(t), \tag{23.6}$$

is well-defined and is a continuous semivalue on $GDIFF_{\mathcal{T}}$ (of norm 1);

(vi) *the restriction of θ to $GDIFFE_{\mathcal{T}}$ (still denoted θ) is a continuous Aumann–Shapley value on $GDIFFE_{\mathcal{T}}$ (of norm 1). $GDIFFE_{\mathcal{T}}$ is the largest subspace of $GDIFF_{\mathcal{T}}$ on which θ is efficient, i.e., such that the semivalue θ is an Aumann–Shapley value.*

The proof of this result will be given in Section 24.

23.7 Remark.

(i) Theorem 23.6 suggests another way of proving some parts of Theorem 18.4. Observing that the games of the form $\boldsymbol{v} = \boldsymbol{m}^k$, where $k \in \mathbb{N}$ and $\boldsymbol{m}$ is a T_∞-measure on the generated tribe $\mathcal{T}$, are contained in $GDIFFE_{\mathcal{T}}$ (cf. Theorem 23.6(iv)) and that $GDIFFE_{\mathcal{T}}$ is closed, it follows that $pFNA_{\mathcal{T}} \subseteq GDIFFE_{\mathcal{T}}$ and, therefore, the restriction of θ to $pFNA_{\mathcal{T}}$ is an Aumann–Shapley value on $pFNA_{\mathcal{T}}$ which necessarily satisfies the diagonal formula (18.6) because of (23.6). However, *by this reasoning, the fact that $\mathcal{T}$ is a generated T_∞-tribe is crucial, and this is not a hypothesis in Theorem 18.4.*

(ii) In Theorem 23.6 formula (23.6) can be restated in the form (20.3) because of the Proposition 23.2. Combined with Remark (i) above, this shows that we can obtain another proof of the part of Theorem 20.4 which states that the diagonal value φ on $pFNA_{\mathcal{T}}$ has the diagonal representation (20.3) if $\mathcal{T}$ is generated. However, the fact that for games $\boldsymbol{v} \in pFNA_{\mathcal{T}}$ the diagonal value $\varphi\boldsymbol{v}$ is a T_∞-measure does not follow from Theorem 23.6. Indeed, if $\boldsymbol{v} \in GDIFFE_{\mathcal{T}}$, then $\varphi\boldsymbol{v}$ may not be a T_∞-measure. For instance, take $\mathcal{T} = [0,1]^{\mathbb{N}}$, i.e., $\mathcal{T}$ is the T_∞-tribe generated by the σ-algebra $\mathcal{A}$ of all crisp subsets of $\mathbb{N}$. Let $\mathcal{F}$ be the filter of the crisp subsets of $\mathbb{N}$ with finite complements, i.e.,

$$\mathcal{F} = \{S \subseteq \mathbb{N} \mid S^c \text{ is finite}\}.$$

Let $\mathcal{U}$ be an ultrafilter such that $\mathcal{F} \subseteq \mathcal{U}$ (recall that such an ultrafilter always exists by the Axiom of Choice). Define $\boldsymbol{m} : \mathcal{A} \longrightarrow \mathbb{R}$ by

$$\boldsymbol{m}(A) = \begin{cases} 1 & \text{if } A \in \mathcal{U}, \\ 0 & \text{if } A \notin \mathcal{U} \end{cases}$$

Note that $\boldsymbol{m}$ is finitely additive but *not a σ-additive* measure on A since if for each $n \in \mathbb{N}$ we choose $A_n = \{n\}$, then $\boldsymbol{m}(A_n) = 0$ while

$$\boldsymbol{m}(\bigcup_{n\in\mathbb{N}} A_m) = \boldsymbol{m}(\mathbb{N}) = 1.$$

As a consequence, the function $\hat{\boldsymbol{m}} : \mathcal{T} \longrightarrow \mathbb{R}$ defined by

$$\hat{\boldsymbol{m}}(A) = \int_X A(x)\, d\boldsymbol{m}(x),$$

where the integral with respect to the finitely additive measure $\boldsymbol{m}$ is defined in the sense of DUNFORD & SCHWARTZ [1958], is *not a T_∞-measure on $\mathcal{T}$*. However, it is obvious that

$$\delta\hat{\boldsymbol{m}}(t, A) = \hat{\boldsymbol{m}}(A)$$

for each $t \in [0,1]$ and $A \in \mathcal{T}$ and, therefore, $\hat{\boldsymbol{m}} \in GDIFFE_{\mathcal{T}}$ and $\theta\hat{\boldsymbol{m}} = \hat{\boldsymbol{m}}$. This shows that there are games $\boldsymbol{v}$ in $GDIFFE_{\mathcal{T}}$ for which $\varphi\boldsymbol{v}$ is not a T_∞-measure. This contrasts with the value of games in $pFNA_{\mathcal{T}}$.

(iii) The fact that for games $\boldsymbol{v} \in pFNA_{\mathscr{T}}$ the diagonal value $\varphi\boldsymbol{v}$ is a T_∞-measure on $\mathscr{T}$ whenever $\mathscr{T}$ is a generated T_∞-tribe is important since it determines the representation (19.4) of the diagonal value φ^* on $pFNA^*_{\mathscr{T}}$ and the corresponding representation of the diagonal value ψ on $pNA_{\mathscr{A}}$ because of $\psi = (\varphi^* \circ e)^\vee$, as shown in Section 19. Note that the representation (22.11) of the Fréchet differential $D\boldsymbol{v}(tX, \cdot)$ is equivalent to the fact that $\varphi\boldsymbol{v}$ is a T_∞-measure. Representation (22.11) was sometimes misinterpreted as saying that the diagonal value $\varphi\boldsymbol{v}$ of a game $\boldsymbol{v} \in pFNA_{\mathscr{T}}$ is uniquely determined by the restriction $\check{\boldsymbol{v}}$ of $\boldsymbol{v}$ to $\mathscr{A}$. The same misinterpretation was sometimes extrapolated to games in $GDIFFE_{\mathscr{T}}$. The truth is that there are games in $pFNA_{\mathscr{T}}$ which have the same restrictions to $\mathscr{A}$ but different values. More precisely, take two T_s-measures $\boldsymbol{m}_1$ and $\boldsymbol{m}_2$ on $\mathscr{T}$ with $s \in]0,\infty[$ such that the restrictions of $\boldsymbol{m}_1$ and $\boldsymbol{m}_2$ to $\mathscr{A}$ yield the same σ-additive measure $\boldsymbol{m}$ on $\mathscr{A}$. Since such functions are representable by formula (7.1), it can be easily seen that for $i \in \{1,2\}$

$$\delta\boldsymbol{m}_i(t, A) = \int_0^1 (1 - f_i) \cdot A \, d\boldsymbol{m},$$

where f_i is the (uniquely determined) function in the representation (7.1) of $\boldsymbol{m}_i$. Clearly $\delta\boldsymbol{m}_i(t, \cdot)$ are T_∞-measures. However, if

$$\int_Y f_1 \, d\boldsymbol{m} \neq \int_Y f_2 \, d\boldsymbol{m}$$

for some crisp set $Y \in \mathscr{A}$, then $\varphi\boldsymbol{m}_1 \neq \varphi\boldsymbol{m}_2$ in spite of $\check{\boldsymbol{m}}_1 = \boldsymbol{m} = \check{\boldsymbol{m}}_2$.

24 Proof of Theorem 23.6

24.1 In this section we prove a sequence of lemmas which finally provide a proof of Theorem 23.6. A proof of a result which is similar to Theorem 23.6 can also be found in [MERTENS 1980]. However, there is a significant difference between Theorem 23.6 and Mertens' corresponding theorem. In fact, Mertens' result is proven for games $\boldsymbol{v}$ in a space we later shall call $MDIFFE_{\mathscr{T}}$, which satisfy the standardness assumption and the condition that the variation $\boldsymbol{v}^* = \boldsymbol{v}^+ + \boldsymbol{v}^-$ has a diagonal function being continuous at the points 0 and 1. This space of games contains $pFNA_{\mathscr{T}}$ but excludes many other games, such as those discussed in Example 23.5 where the diagonal function f is monotone but discontinuous at one or at both endpoints of the interval $[0, 1]$. Also, the proof of Theorem 23.6 we present below is, in principle, based on the properties of Dini derivatives proven in this work.

24.2 Note that $GDIFFE_{\mathscr{T}}$ is a proper subset of $GDIFF_{\mathscr{T}}$ (see Example 23.5). Therefore, the following result proves Theorem 23.6(i, ii).

Lemma. *$GDIFF_{\mathscr{T}}$ and $GDIFFE_{\mathscr{T}}$ are symmetric closed subspaces of $FBV_{\mathscr{T}}$. For each $\boldsymbol{v} \in FBV_{\mathscr{T}}$, for each automorphism ξ of the measure space $(X, \mathscr{A})$ and for all $A \in \mathscr{T}$ we have*

$$\delta(\xi_* \boldsymbol{v})(t, A) = \delta \boldsymbol{v}(t, \xi A). \tag{24.1}$$

PROOF: It is clear that $GDIFF_{\mathscr{T}}$ and $GDIFFE_{\mathscr{T}}$ are linear subspaces of $FBV_{\mathscr{T}}$. Let $\boldsymbol{v} \in GDIFF_{\mathscr{T}}$ and let ξ be an automorphism of $(X, \mathscr{A})$. For each $t \in [0,1[$ and for each $\tau \in]0, 1-t[$ we have

$$\frac{(\xi_* \boldsymbol{v})(tX + \tau A) - (\xi_* \boldsymbol{v})(tX)}{\tau} = \frac{\boldsymbol{v}(tX + \tau \xi A) - \boldsymbol{v}(tX)}{\tau},$$

showing that (24.1) holds and, implicitly, that $GDIFF_{\mathscr{T}}$ is symmetric. Since $g_{\xi_* \boldsymbol{v}} = g_{\boldsymbol{v}}$, it follows that $GDIFFE_{\mathscr{T}}$ is symmetric too. In order to prove that $GDIFF_{\mathscr{T}}$ is closed, let $(\boldsymbol{v}_n)_{n \in \mathbb{N}}$ be a sequence in $GDIFF_{\mathscr{T}}$ which converges to $\boldsymbol{v} \in FBV_{\mathscr{T}}$. By Lemma 21.8, there exists a set $M \subseteq [0,1]$ such that $\lambda(M) = 1$ and such that for each $t \in M$ and all $n \in \mathbb{N}$ the Dini derivative $\delta \boldsymbol{v}_n(t, \cdot)$ exists and is finite. For each $t \in M$ and each $A \in \mathscr{T}$ we have

$$\begin{aligned} |\delta \boldsymbol{v}_n(t, A) - \delta \boldsymbol{v}_m(t, A)| &= |\delta(\boldsymbol{v}_n - \boldsymbol{v}_m)(t, A)| \\ &= \partial^+(\boldsymbol{v}_n - \boldsymbol{v}_m)(t, A) \\ &\leq \Delta^+(\boldsymbol{v}_n - \boldsymbol{v}_m)(t). \end{aligned}$$

Applying Lemma 21.5 we obtain

$$\begin{aligned} \int_0^1 \sup_{A \in \mathscr{T}} \{|\delta \boldsymbol{v}_n(t, A) - \delta \boldsymbol{v}_m(t, A)|\} \, \mathrm{d}\lambda(t) &= \int_0^1 \sup_{A \in \mathscr{T}} \{|\delta(\boldsymbol{v}_n - \boldsymbol{v}_m)(t, A)|\} \, \mathrm{d}\lambda(t) \\ &\leq \int_0^1 \Delta^+(\boldsymbol{v}_n - \boldsymbol{v}_m)(t) \, \mathrm{d}\lambda(t) \\ &\leq \|\boldsymbol{v}_n - \boldsymbol{v}_m\|, \end{aligned}$$

for all $m, n \in \mathbb{N}$. This and Lemma 24.1 imply that the sequence of functions $(\delta \boldsymbol{v}_n(\cdot, A))_{n \in \mathbb{N}}$ is fundamental in the Lebesgue measure uniformly with respect to A. Theorem A of HALMOS [1974, p. 93] implies that the sequence $(\delta \boldsymbol{v}_n(\cdot, A))_{n \in \mathbb{N}}$ has a λ-almost uniformly convergent subsequence (which converges uniformly with respect to A on a set of Lebesgue measure 1). We still denote this subsequence $(\delta \boldsymbol{v}_n(\cdot, A))_{n \in \mathbb{N}}$, and we denote by $\delta' \boldsymbol{v}(\cdot, A)$ its λ-a.e. limit. Clearly, $\delta' \boldsymbol{v}(\cdot, A)$ is Borel measurable. We claim that for λ-almost all $t \in [0,1]$, $\delta' \boldsymbol{v}(t, A)$ equals the Dini derivative of the game $\boldsymbol{v}$ at tX in the direction A. In order to prove this, observe that

$$\int_0^1 \limsup_{\tau \searrow 0} \left| \delta' \boldsymbol{v}(t, A) - \frac{\boldsymbol{v}(tX \, \mathbf{S}_\infty \, \tau A) - \boldsymbol{v}(tX)}{\tau} \right| \mathrm{d}\lambda(t) \leq$$

$$\le \int_0^1 |\delta' \boldsymbol{v}(t,A) - \delta \boldsymbol{v}_n(t,A)| \, \mathrm{d}\lambda(t)$$

$$+ \int_0^1 \limsup_{\tau \searrow 0} \left| \delta \boldsymbol{v}(t,A) - \frac{\boldsymbol{v}(tX \mathbf{S}_\infty \tau A) - \boldsymbol{v}(tX)}{\tau} \right| \mathrm{d}\lambda(t)$$

$$\le \int_0^1 |\delta' \boldsymbol{v}(t,A) - \delta \boldsymbol{v}_n(t,A)| \, \mathrm{d}\lambda(t)$$

$$+ \int_0^1 \limsup_{\tau \searrow 0} \left| \delta \boldsymbol{v}_n(t,A) - \frac{\boldsymbol{v}_n(tX \mathbf{S}_\infty \tau A) - \boldsymbol{v}_n(tX)}{\tau} \right| \mathrm{d}\lambda(t)$$

$$+ \int_0^1 \limsup_{\tau \searrow 0} \left| \frac{\boldsymbol{v}_n(tX \mathbf{S}_\infty \tau A) - \boldsymbol{v}_n(tX)}{\tau} - \frac{\boldsymbol{v}(tX \mathbf{S}_\infty \tau A) - \boldsymbol{v}(tX)}{\tau} \right| \mathrm{d}\lambda(t).$$

For λ-almost all $t \in [0,1]$ and for all $n \in \mathbb{N}$ we get

$$\limsup_{\tau \searrow 0} \left| \delta \boldsymbol{v}_n(t,A) - \frac{\boldsymbol{v}_n(tX \mathbf{S}_\infty \tau A) - \boldsymbol{v}_n(tX)}{\tau} \right| = 0.$$

Also, note that for each $\varepsilon > 0$ there exists a number $n(\varepsilon) \in \mathbb{N}$ such that for λ-almost all $t \in [0,1]$ we have

$$|\delta' \boldsymbol{v}(t,A) - \delta \boldsymbol{v}_n(t,A)| \le \varepsilon$$

whenever $n \ge n(\varepsilon)$. Therefore, for each $\varepsilon > 0$ there exists a number $n(\varepsilon) \in \mathbb{N}$ such that for all $n \ge n(\varepsilon)$ we have

$$\int_0^1 \limsup_{\tau \searrow 0} \left| \delta' \boldsymbol{v}(t,A) - \frac{\boldsymbol{v}(tX \mathbf{S}_\infty \tau A) - \boldsymbol{v}(tX)}{\tau} \right| \mathrm{d}\lambda(t) \le$$

$$\le \varepsilon + \int_0^1 \limsup_{\tau \searrow 0} \left| \frac{\boldsymbol{v}_n(tX \mathbf{S}_\infty \tau A) - \boldsymbol{v}_n(tX)}{\tau} - \frac{\boldsymbol{v}(tX \mathbf{S}_\infty \tau A) - \boldsymbol{v}(tX)}{\tau} \right| \mathrm{d}\lambda(t)$$

$$\le \varepsilon + \int_0^1 \limsup_{\tau \searrow 0} \left| \frac{(\boldsymbol{v}_n - \boldsymbol{v})(tX \mathbf{S}_\infty \tau A) - (\boldsymbol{v}_n - \boldsymbol{v})(tX)}{\tau} \right| \mathrm{d}\lambda(t)$$

$$= \varepsilon + \int_0^1 |\partial^+ (\boldsymbol{v}_n - \boldsymbol{v})(t,A)| \, \mathrm{d}\lambda(t).$$

By Lemma 21.5 for each $\varepsilon > 0$ there exists a number $n(\varepsilon) \in \mathbb{N}$ such that for each $n \ge n(\varepsilon)$

$$\int_0^1 \limsup_{\tau \searrow 0} \left| \delta' \boldsymbol{v}(t,A) - \frac{\boldsymbol{v}(tX \mathbf{S}_\infty \tau A) - \boldsymbol{v}(tX)}{\tau} \right| \mathrm{d}\lambda(t) \le \varepsilon + \|\boldsymbol{v}_n - \boldsymbol{v}\|$$

or, equivalently,

$$\int_0^1 \limsup_{\tau \searrow 0} \left| \delta' \boldsymbol{v}(t,A) - \frac{\boldsymbol{v}(tX \mathbf{S}_\infty \tau A) - \boldsymbol{v}(tX)}{\tau} \right| \mathrm{d}\lambda(t) = 0.$$

Thus for λ-almost all $t \in [0,1]$

$$\begin{aligned} 0 &\le \liminf_{\tau \searrow 0} \left| \delta' \boldsymbol{v}(t,A) - \frac{\boldsymbol{v}(tX\,\mathbf{S}_\infty\,\tau A) - \boldsymbol{v}(tX)}{\tau} \right| \\ &\le \limsup_{\tau \searrow 0} \left| \delta' \boldsymbol{v}(t,A) - \frac{\boldsymbol{v}(tX\,\mathbf{S}_\infty\,\tau A) - \boldsymbol{v}(tX)}{\tau} \right| \\ &= 0. \end{aligned}$$

Hence, for λ-almost all $t \in [0,1]$, $\delta\boldsymbol{v}(t,A)$ exists and $\delta'\boldsymbol{v}(t,A) = \delta\boldsymbol{v}(t,A)$. Now, if A and B are T_∞-disjoint fuzzy sets in $\mathcal{T}$ and if $t \in [0,1]$ is a point at which $\delta\boldsymbol{v}(t,A)$ and $\delta\boldsymbol{v}(t,B)$ exist, then $\delta\boldsymbol{v}(t, A\,\mathbf{S}_\infty\, B)$ also exists since

$$\lim_{n\to\infty} \delta\boldsymbol{v}_n(t,A) + \lim_{n\to\infty} \delta\boldsymbol{v}_n(t,B) = \lim_{n\to\infty} \delta\boldsymbol{v}_n(t, A\,\mathbf{S}_\infty\, B) = \delta\boldsymbol{v}(t, A\,\mathbf{S}_\infty\, B).$$

This shows that $\delta\boldsymbol{v}(t,\cdot)$ is finitely T_∞-additive. Hence, $\boldsymbol{v} \in \mathit{GDIFF}_{\mathcal{T}}$ and this proves that $\mathit{GDIFF}_{\mathcal{T}}$ is closed. Now suppose that the sequence $(\boldsymbol{v}_n)_{n\in\mathbb{N}}$ is contained in $\mathit{GDIFFE}_{\mathcal{T}}$. Then for each $n \in \mathbb{N}$ we have

$$\begin{aligned} \int_0^1 |g'_{\boldsymbol{v}_n}(t) - g'_{\boldsymbol{v}}(t)|\,\mathrm{d}\lambda(t) &= \int_0^1 |\delta(\boldsymbol{v} - \boldsymbol{v}_n)(t,X)|\,\mathrm{d}\lambda(t) \\ &\le \fint_0^1 \partial^+(\boldsymbol{v} - \boldsymbol{v}_n)(t,X)\,\mathrm{d}\lambda(t) \\ &\le \|\boldsymbol{v}_n - \boldsymbol{v}\|, \end{aligned}$$

where $\|\boldsymbol{v}_n - \boldsymbol{v}\| \to 0$ if $n \to \infty$. Therefore, we get

$$\int_0^1 g'_{\boldsymbol{v}}(t)\,\mathrm{d}\lambda(t) = \lim_{n\to\infty} g'_{\boldsymbol{v}_n}(t)\,\mathrm{d}\lambda(t) = \lim_{n\to\infty} \boldsymbol{v}_n(X) = \boldsymbol{v}(X),$$

i.e., $\boldsymbol{v} \in \mathit{GDIFFE}_{\mathcal{T}}$, and this implies that $\mathit{GDIFFE}_{\mathcal{T}}$ is closed. □

24.3 Taking into account Lemma 24.2, the following lemma will prove Theorem 23.6(iii). Note especially that the games considered there are not necessarily diagonally efficient (see Example 23.5).

Lemma. *The games of the form $\boldsymbol{v} = f \circ \boldsymbol{m}$, where $\boldsymbol{m}$ is a probability T_∞-measure on $\mathcal{T}$ and $f : [0,1] \longrightarrow \mathbb{R}$ is a function of bounded variation such that $f(0) = 0$, belong to $\mathit{GDIFF}_{\mathcal{T}}$.*

PROOF: Since the tribe $\mathcal{T}$ is generated, $\boldsymbol{m}$ is necessarily nonatomic on $\mathcal{T}$ and, therefore, $\mathcal{R}(\boldsymbol{m}) = [0,1]$. Consequently, $\|\boldsymbol{v}\|$ is less than or equal to the variation of f. Therefore, $\boldsymbol{v} \in \mathit{FBV}_{\mathcal{T}}$. Note that for each $A \in \mathcal{T}$

$$\delta\boldsymbol{v}(t,A) = f'(t)\cdot \boldsymbol{m}(A) \tag{24.2}$$

whenever f is differentiable at t. Hence, $\boldsymbol{v}$ is G-differentiable. □

24.4 Theorem 23.6 (iv) is an immediate consequence of Lemma 24.3 and of the following result, which was first proved by H. Lebesgue in 1903 (see [HAWKINS 1970, p. 130] and [NATANSON 1969, p. 301]).

Lemma. *Let $f : [0,1] \longrightarrow \mathbf{R}$ be a function having finite derivative $f'(t)$ at each point $t \in [0,1]$. The function f' is λ-integrable if and only if f is of bounded variation. If f' is integrable, then*

$$\int_0^1 f'(t)\,\mathrm{d}\lambda(t) = f(1) - f(0). \tag{24.3}$$

24.5 Now we are in the position to prove Theorem 23.6(v, vi) and, therefore, to complete the proof of this theorem.

Lemma. *The operator $\theta : GDIFF_{\mathscr{T}} \longrightarrow FBA_{\mathscr{T}}$ given by (23.6) is well-defined. It is a continuous semivalue (of norm 1). The restriction of θ to the space $GDIFFE_{\mathscr{T}}$ is an Aumann–Shapley value. $GDIFFE_{\mathscr{T}}$ is the largest subspace of $GDIFF_{\mathscr{T}}$ on which θ is an Aumann–Shapley value.*

PROOF: By Proposition 23.2 the integral in (23.6) exists, and from Lemma 21.5 it follows that

$$|\theta\boldsymbol{v}(A)| \leq \|\boldsymbol{v}\| \tag{24.4}$$

for all $\boldsymbol{v} \in GDIFF_{\mathscr{T}}$ and for all $A \in \mathscr{T}$. Since for a game $\boldsymbol{v}$ in $GDIFF_{\mathscr{T}}$ the Dini derivative $\delta\boldsymbol{v}(t,\cdot)$ is necessarily finitely T_∞-additive, it follows that for each $\boldsymbol{v} \in GDIFF_{\mathscr{T}}$ the function $\theta\boldsymbol{v}$ is finitely T_∞-additive on $\mathscr{T}$. Hence, θ is well-defined. It is clear that θ is linear. From (24.4) we deduce that the variation $(\theta\boldsymbol{v})^*$ of $\theta\boldsymbol{v}$ (see Definition 9.2) is bounded and that

$$\|\theta\boldsymbol{v}\| = (\theta\boldsymbol{v})^*(X) \leq \|\boldsymbol{v}\|$$

whenever $\boldsymbol{v} \in GDIFF_{\mathscr{T}}$. Consequently, θ is continuous, its norm being less than or equal to 1. On the other hand, note that, according to Lemma 24.2, finite T_∞-measures are contained in $GDIFF_{\mathscr{T}}$ and if $\boldsymbol{m}$ is a probability T_∞-measure, then $\|\theta\boldsymbol{m}\| = \|\boldsymbol{m}\| = \boldsymbol{m}(X) = 1$ showing that the norm of θ is greater than or equal to 1. Hence, θ is continuous, its norm equaling 1. By Lemma 21.7 the operator θ is positive. It is also symmetric because of (24.1). Hence, θ is a semivalue on $GDIFF_{\mathscr{T}}$. If $\boldsymbol{v} \in GDIFFE_{\mathscr{T}}$, then

$$\theta\boldsymbol{v}(X) = \int_0^1 \delta\boldsymbol{v}(t, X)\,\mathrm{d}\lambda(t) = \int_0^1 g_{\boldsymbol{v}}'(t)\,\mathrm{d}\lambda(t) = \boldsymbol{v}(X), \tag{24.5}$$

showing that θ is a value on $GDIFFE_{\mathscr{T}}$. If Q is another symmetric subspace of $GDIFF_{\mathscr{T}}$ on which θ is an Aumann–Shapley value, then for each game $\boldsymbol{v} \in Q$ equation (24.5) is necessarily satisfied and, therefore, Q is a subspace of $GDIFFE_{\mathscr{T}}$. □

25 Further Extensions of the Diagonal Value

25.1 The constructions in Sections 20-24 give rise to the question whether $GDIFFE_{\mathscr{T}}$ is a maximal subspace of $FBV_{\mathscr{T}}$ on which a diagonal Aumann–Shapley value exists. We shall see in this section that this is not the case, i.e., $GDIFFE_{\mathscr{T}}$ is not a maximal subspace with this property. In fact, we shall show that the diagonal value φ on $pFNA_{\mathscr{T}}$ can be extended continuously in various ways to different spaces of games with bounded variation. Let us first recall the operator we denote ξ and which was introduced by MERTENS [1980].

Definition. Let $\mathscr{T}$ be a T_∞-tribe on X.

(i) A game $\boldsymbol{v} : \mathscr{T} \longrightarrow \mathbb{R}$ is said to be *differentiable in the sense of Mertens* (or, briefly, *M-differentiable*) if for each $A \in \mathscr{T}$ the following limit exists and is finite:

$$\xi\boldsymbol{v}(A) = \lim_{\tau \searrow 0} \int_0^1 \frac{\boldsymbol{v}(tX \mathbf{S}_\infty \tau A) - \boldsymbol{v}(tX)}{\tau} \, \mathrm{d}\lambda(t). \tag{25.1}$$

(ii) The set of all M-differentiable games in $FBV_{\mathscr{T}}$ is denoted $MDIFF_{\mathscr{T}}$; the subset of all games $\boldsymbol{v} \in MDIFF_{\mathscr{T}}$ such that $\xi\boldsymbol{v}(X) = \boldsymbol{v}(X)$ will be denoted $MDIFFE_{\mathscr{T}}$.

MERTENS [1980] proved that $MDIFF_{\mathscr{T}}$ and $MDIFFE_{\mathscr{T}}$ are symmetric closed subspaces of $FBV_{\mathscr{T}}$, that ξ is an Aumann–Shapley value on $MDIFFE_{\mathscr{T}}$ and that $MDIFFE_{\mathscr{T}}$ contains significant classes of games. Implicitly, he also showed that $pFNA_{\mathscr{T}} \subseteq MDIFFE_{\mathscr{T}}$ and that ξ is a continuous extension of the Aumann–Shapley value φ on $pFNA_{\mathscr{T}}$.

25.2 Obviously, there are games in $MDIFFE_{\mathscr{T}}$ which do not belong to $GDIFFE_{\mathscr{T}}$, i.e., $MDIFFE_{\mathscr{T}}$ is not a subspace of $GDIFFE_{\mathscr{T}}$. For instance, the game $\boldsymbol{v} = f \circ \hat{\lambda}$ defined in Example 23.5, where f is given by (23.5) and by (23.3), is an element of $MDIFFE_{\mathscr{T}}$ without belonging to $GDIFFE_{\mathscr{T}}$. However, the space $GDIFFE_{\mathscr{T}}$ is not a subspace of $MDIFFE_{\mathscr{T}}$ either. Here is an example of a game in $GDIFFE_{\mathscr{T}}$ which does not belong to $MDIFFE_{\mathscr{T}}$.

Example. Let $\mathscr{B}$ be the σ-algebra of all Borel subsets of $[0,1]$ and $\mathscr{T} = \mathscr{B}^\wedge$. Consider the game $\boldsymbol{v} = f \circ \hat{\lambda}$, where the T_∞-measure on $\mathscr{T}$ is given by (23.3) and the step function $f : [0,1] \longrightarrow [0,1]$ is defined as follows:

$$f(t) = \begin{cases} 1 & \text{if } t \in]0, 0.5], \\ 0 & \text{otherwise.} \end{cases}$$

Clearly, f is of bounded variation for it may be written as $f = f_1 - f_2$, the monotone functions $f_1, f_2 : [0,1] \longrightarrow [0,1]$ being given by

$$f_1(t) = \begin{cases} 1 & \text{if } t \in]0, 1], \\ 0 & \text{otherwise,} \end{cases}$$

$$f_2(t) = \begin{cases} 1 & \text{if } t \in]0.5, 1], \\ 0 & \text{otherwise.} \end{cases}$$

This implies that $\boldsymbol{v} \in FBV_{\mathscr{T}}$. Note that $\boldsymbol{v} \in GDIFFE_{\mathscr{T}}$ and $\theta\boldsymbol{v}(A) = 0$ for all $A \in \mathscr{T}$. On the other hand, we have $\boldsymbol{v}(tX) = f(t)$ and

$$\boldsymbol{v}(tX \, \mathbf{S}_\infty \, \tau A) = \begin{cases} 1 & \text{if } t + \tau \cdot \hat{\lambda}(A) \leq 0.5, \\ 0 & \text{otherwise} \end{cases}$$

for all $(t, \tau) \in [0,1] \times]0,1]$. Therefore, we obtain

$$\xi\boldsymbol{v}(A) = \lim_{\tau \searrow 0} \int_{-0.5-\tau\cdot\hat{\lambda}(A)}^{0.5} (-\frac{1}{\tau}) \, \mathrm{d}\lambda(t) = -\hat{\lambda}(A)$$

which yields

$$\xi\boldsymbol{v}(X) = -1 \neq 0 = \theta\boldsymbol{v}(X),$$

i.e., $\boldsymbol{v} \notin MDIFFE_{\mathscr{T}}$.

25.3 The existence of substantially different (continuous) extensions of the diagonal Aumann–Shapley value φ on $pFNA_{\mathscr{T}}$ seems to be intimately related to the nonstrict *convexity* of the space $pFNA_{\mathscr{T}}$ (since $\|\boldsymbol{u}\| = \|\boldsymbol{w}\| = \frac{1}{2}\|(\boldsymbol{u} + \boldsymbol{w})\|$ whenever $\boldsymbol{u}$ and $\boldsymbol{w}$ are probability T_∞-measures on $\mathscr{T}$). Moreover, each of the Aumann–Shapley values θ and ξ which exists on $GDIFFE_{\mathscr{T}}$ and $MDIFFE_{\mathscr{T}}$, respectively, can be further extended continuously to Auman–Shapley values on significantly larger subspaces of $FBV_{\mathscr{T}}$, none of which are maximal in the sense that no Aumann–Shapley value can be defined on spaces of which they are proper subspaces.

The fact that the Aumann–Shapley value ξ on $MDIFFE_{\mathscr{T}}$ can be further extended to a subspace of $FBV_{\mathscr{T}}$ which is strictly larger than $MDIFFE_{\mathscr{T}}$, follows from the work of MERTENS [1988]. The basic feature of this extension is the following operator ξ^*:

Definition. Let $\mathscr{T}$ be a T_∞-tribe on X. For a game $\boldsymbol{v} : \mathscr{T} \longrightarrow \mathbb{R}$ define

$$\xi^* \boldsymbol{v}(A) = \lim_{\tau \searrow 0} \int_0^1 \frac{\boldsymbol{v}(tX + \tau A) - \boldsymbol{v}(tX - \tau A)}{2\tau} \, \mathrm{d}\lambda(t). \tag{25.2}$$

The set of games $\boldsymbol{v} \in FBV_{\mathscr{T}}$ such that the limit exists and is finite for all $A \in \mathscr{T}$ and such that the function $\xi^* \boldsymbol{v} : \mathscr{T} \longrightarrow \mathbb{R}$ is finitely T_∞-additive is denoted here $MDIFF^*_{\mathscr{T}}$. The subset of all games $\boldsymbol{v} \in MDIFF^*_{\mathscr{T}}$ such that $\xi^* \boldsymbol{v}(X) = \boldsymbol{v}(X)$ will be denoted $MDIFFE^*_{\mathscr{T}}$. It is clear that $MDIFF_{\mathscr{T}} \subseteq MDIFF^*_{\mathscr{T}}$.

MERTENS [1988] proved that $MDIFFE^*_{\mathscr{T}}$ is a symmetric closed subspace of $FBV_{\mathscr{T}}$ and that ξ^* is a continuous Aumann–Shapley value on this space. $MDIFFE^*_{\mathscr{T}}$ is strictly larger than $MDIFFE_{\mathscr{T}}$ since, for instance, it contains the game of Example 25.2 which we have seen is not in $MDIFFE_{\mathscr{T}}$.

25.4 There is yet another way of further extending the Aumann–Shapley value φ on $pFNA_{\mathscr{T}}$. It consists of an extension of the Aumann–Shapley value θ on $GDIFFE_{\mathscr{T}}$. In order to discuss this extension we introduce the following notation:

Definition. The set $GDIFF^*_{\mathscr{T}}$ consists of all games $\boldsymbol{v} \in FBV_{\mathscr{T}}$ such that

(i) for λ-almost all $t \in [0,1]$ and for all $A \in \mathscr{T}$ the following limit exists:

$$\gamma \boldsymbol{v}(t, A) = \lim_{\tau \searrow 0} \frac{\boldsymbol{v}(tX + \tau A) - \boldsymbol{v}(tX - \tau A)}{2\tau}; \tag{25.3}$$

(ii) for each $A \in \mathscr{T}$ the function $\gamma \boldsymbol{v}(\cdot, A)$ given below is λ-integrable:

$$\gamma \boldsymbol{v}(\cdot, A) : [0,1] \longrightarrow [-\infty, +\infty], \quad t \longmapsto \gamma \boldsymbol{v}(t, A);$$

(iii) for λ-almost all $t \in [0,1]$ the function $\gamma \boldsymbol{v}(t, \cdot)$ given below is finitely T_∞-additive:

$$\gamma \boldsymbol{v}(t, \cdot) : \mathscr{T} \longrightarrow [-\infty, +\infty], \quad A \longmapsto \gamma \boldsymbol{v}(t, A).$$

The subset of all games $\boldsymbol{v} \in GDIFF^*_{\mathscr{T}}$ being diagonally efficient is denoted $GDIFFE^*_{\mathscr{T}}$.

At this moment we do not know whether condition (ii) in the definition above follows from (i) and some other properties of the games in $FBV_{\mathscr{T}}$.

25.5 It is obvious that $GDIFFE_{\mathscr{T}} \subseteq GDIFFE^*_{\mathscr{T}}$. Moreover, $GDIFFE_{\mathscr{T}}$ is a proper subspace of $GDIFFE^*_{\mathscr{T}}$ since there are games in $GDIFFE^*_{\mathscr{T}}$ which do not belong to $GDIFFE_{\mathscr{T}}$. Probably the most interesting examples among them are the so-called *2-handed glove market games* as defined below (for more details about the class of n-handed glove market games see [AUMANN & SHAPLEY 1974], [HART 1980], [TAUMANN 1981] and [MONDERER 1989]).

Example. Let $\mathscr{A}$ be a σ-algebra of crisp subsets of X and put $\mathscr{T} = \mathscr{A}^\wedge$. Consider n probability measures $\boldsymbol{m}_1, \boldsymbol{m}_2, \ldots, \boldsymbol{m}_n$ on $\mathscr{A}$ and for each $i \in \{1, 2, \ldots, n\}$ define $\hat{\boldsymbol{m}}_i : \mathscr{T} \longrightarrow \mathbb{R}$ by

$$\hat{\boldsymbol{m}}_i(A) = \int_X A(x)\,\mathrm{d}\boldsymbol{m}_i(x).$$

The *n-handed glove market game* $\boldsymbol{v} : \mathscr{T} \longrightarrow \mathbb{R}$ associated with the probabilities $\boldsymbol{m}_1, \boldsymbol{m}_2, \ldots, \boldsymbol{m}_n$ is given by

$$\boldsymbol{v}(A) = \min(\hat{\boldsymbol{m}}_1(A), \hat{\boldsymbol{m}}_2(A), \ldots, \hat{\boldsymbol{m}}_n(A)).$$

It is clear that $\boldsymbol{v} \in FBV^+_{\mathscr{T}}$ and that for all $(t, A) \in [0,1] \times \mathscr{T}$ we have $\delta\boldsymbol{v}(t, A) = \boldsymbol{v}(A)$ which means that $\boldsymbol{v}$ is Dini derivable.

For $n \geq 2$, however, $\boldsymbol{v}$ is not Gâteaux–differentiable since the function $\delta\boldsymbol{v}(t, \cdot)$ is not finitely T_∞-additive and, therefore, $\boldsymbol{v} \notin GDIFFE_{\mathscr{T}}$.

It is easy to see that we have

$$\gamma\boldsymbol{v}(t, A) = \frac{1}{2} \cdot [\min(\hat{\boldsymbol{m}}_1(A), \ldots, \hat{\boldsymbol{m}}_n(A)) + \max(\hat{\boldsymbol{m}}_1(A), \ldots, \hat{\boldsymbol{m}}_n(A))]$$

for all $(t, A) \in [0,1] \times \mathscr{T}$. In the case $n = 2$ this gives

$$\gamma\boldsymbol{v}(t, A) = \frac{1}{2} \cdot (\hat{\boldsymbol{m}}_1(A) + \hat{\boldsymbol{m}}_2(A))$$

showing that the 2-handed glove market game belongs to $GDIFFE^*_{\mathscr{T}}$.

25.6 The following result shows that the diagonal Aumann–Shapley value θ on $GDIFFE_{\mathscr{T}}$ can be extended continuously to $GDIFFE^*_{\mathscr{T}}$.

Theorem. *Let $\mathscr{T}$ be a generated T_∞-tribe on X. Then we have:*

(i) $GDIFFE^*_{\mathscr{T}}$ is a closed symmetric subspace of $FBV_{\mathscr{T}}$;

(ii) $GDIFFE^*_{\mathscr{T}}$ is a closed symmetric subspace of $GDIFF^*_{\mathscr{T}}$ and contains $GDIFFE_{\mathscr{T}}$ as a proper subspace;

(iii) the function $\theta^* : GDIFFE^*_{\mathscr{T}} \longrightarrow FBA_{\mathscr{T}}$, $\boldsymbol{v} \longmapsto \theta^*\boldsymbol{v}$, given by

$$\theta^*\boldsymbol{v}(A) = \int_0^1 \gamma\boldsymbol{v}(t, A)\,\mathrm{d}\lambda(t) \tag{25.4}$$

is a well-defined semivalue on $GDIFF^*_{\mathscr{T}}$; its restriction to $GDIFFE^*_{\mathscr{T}}$ is an Aumann–Shapley value which is an extension of the Aumann–Shapley value θ on $GDIFFE_{\mathscr{T}}$;

(iv) the Aumann–Shapley value θ^* on $GDIFFE^*_{\mathscr{T}}$ is continuous and has norm 1.

PROOF: We may follow the pattern of the proof of Theorem 23.6 using the differential operator γ instead of the Dini derivative δ (which may not exist for games in $GDIFFE^*_{\mathcal{T}}$). For the sake of brevity we refrain from merely copying Section 24.

The basic technical observation, before working through the several steps of the proof, is that the differential operator γ satisfies

$$\gamma \boldsymbol{v}(t, A) \leq \frac{1}{2} \cdot [\delta^+ \boldsymbol{v}(t, A) + \delta^+ \boldsymbol{v}(t, A^c)] \tag{25.5}$$

since for each sufficiently small $\tau > 0$ we have

$$\frac{\boldsymbol{v}(tX + \tau A) - \boldsymbol{v}(tX - \tau A)}{2\tau}$$
$$= \frac{1}{2} \cdot \left[\frac{\bar{h}_{\boldsymbol{v},A}(t, \tau) - \bar{h}_{\boldsymbol{v},A}(t, 0)}{\tau} - \frac{\bar{h}_{\boldsymbol{v},A^c}(t, \tau) - \bar{h}_{\boldsymbol{v},A^c}(t, 0)}{\tau} \right]$$

with $h_{\boldsymbol{v},A} : \mathbb{R}^2 \longrightarrow \mathbb{R}$ being the function defined in (21.1) and $\bar{h}_{\boldsymbol{v},A} = h_{\boldsymbol{v},A} \circ h$, where the nonsingular transformation $h : \mathbb{R}^2 \longrightarrow \mathbb{R}^2$ is given by $h(t, \tau) = (t - \tau, \tau)$. Now it is a matter of elementary computation to deduce that $\gamma \boldsymbol{v}(\cdot, A)$ is finite λ-a.e. whenever $\boldsymbol{v} \in FBV_{\mathcal{T}}$ and $A \in \mathcal{T}$. Moreover,

$$\int_0^1 |\gamma \boldsymbol{v}(t, A)| \, \mathrm{d}\lambda(t) \leq \fint_0^1 \Delta^+ \boldsymbol{v}(t) \, \mathrm{d}\lambda(t) \leq \|\boldsymbol{v}\|$$

because of Lemmas 21.8 and 21.5.

This provides the series of inequalities needed in order to show that $GDIFF^*_{\mathcal{T}}$ and $GDIFFE^*_{\mathcal{T}}$ are closed and that θ^* is continuous of norm 1. □

25.7 Comparing Theorem 25.6 with the existence theorem for the Aumann–Shapley value ξ^* on $MDIFFE^*_{\mathcal{T}}$ proved by MERTENS [1988], it turns out that in our result the bounded variation is essential for the existence of θ^*. This is in contrast to Mertens' result concerning the existence of ξ^* which can be extrapolated to games without bounded variation.

Moreover, after a little bit of playing around, an example similar to that in 25.1 can be produced showing that the spaces $GDIFFE^*_{\mathcal{T}}$ and $MDIFFE^*_{\mathcal{T}}$ are also incomparable, i.e., neither one is a subspace of the other. The reason behind this is that, in general, the order of limit and the integral in formula (25.2) cannot be switched without changing the value of ξ^*. However, $pFNA_{\mathcal{T}}$ is a proper subspace of the intersection of $GDIFFE^*_{\mathcal{T}}$ and $MDIFFE^*_{\mathcal{T}}$ since the latter contains all 2-handed glove market games. Moreover, it is easy to see that whenever $\boldsymbol{v} \in pFNA_{\mathcal{T}}$, we have

$$\theta^* \boldsymbol{v} = \theta \boldsymbol{v} = \xi \boldsymbol{v} = \xi^* \boldsymbol{v}.$$

But at this moment it is not clear to us whether we really have the equality

$$pFNA_{\mathcal{T}} = GDIFFE^*_{\mathcal{T}} \cap MDIFFE^*_{\mathcal{T}},$$

since we have neither a proof for it nor a counterexample of a game $\boldsymbol{v} \in GDIFFE^*_{\mathcal{T}} \cap MDIFFE^*_{\mathcal{T}}$ such that $\theta^* \boldsymbol{v} \neq \xi^* \boldsymbol{v}$.

Space	Value	Semivalue	Reference
pNA	ψ		AUMANN & SHAPLEY [1974]
$pFNA^*_{\mathcal{T}}$	φ^*		AUMANN & SHAPLEY [1974]
$MDIFF_{\mathcal{T}}$	none	ξ	MERTENS [1980]
$MDIFFE_{\mathcal{T}}$	ξ		MERTENS [1980]
$pFNA_{\mathcal{T}}$	φ		BUTNARIU [1987]
$MDIFF^*_{\mathcal{T}}$	none	ξ^*	MERTENS [1988]
$MDIFFE^*_{\mathcal{T}}$	ξ^*		MERTENS [1988]
$GDIFF_{\mathcal{T}}$	none	θ	Section 23
$GDIFFE_{\mathcal{T}}$	θ		Section 23
$GDIFF^*_{\mathcal{T}}$	none	θ^*	Section 25
$GDIFFE^*_{\mathcal{T}}$	θ^*		Section 25

Table 25.1: Spaces of games with Aumann–Shapley (semi-)values

25.8 Table 25.1 summarizes our discussion about the existence of diagonal Aumann–Shapley (semi-) values on various spaces of games with fuzzy coalitions

The inclusions between the various spaces mentioned in Table 25.1 are shown in Figure 25.1. Table 25.1 and Figure 25.1 deal with axiomatically defined values only, and they will be somewhat enriched in Section 26.

The question arises where *asymptotic* and *mixing values* [AUMANN & SHAPLEY 1974], which are not known to have diagonal representations by differential operators, fit into this picture. The answer is obvious for asymptotic values on subspaces of *pNA* considered by MONDERER [1986].

An analysis of the proofs above gives some hope that the Aumann–Shapley values on *bv'NA*, *MIX* and *ASYMP* may also have diagonal representations for their values, although this cannot be derived directly from our results.

The relevance of finding such diagonal representations for Aumann–Shapley values comes from the fact that such representations provide additional information about the value and enhance its transferability to other spaces.

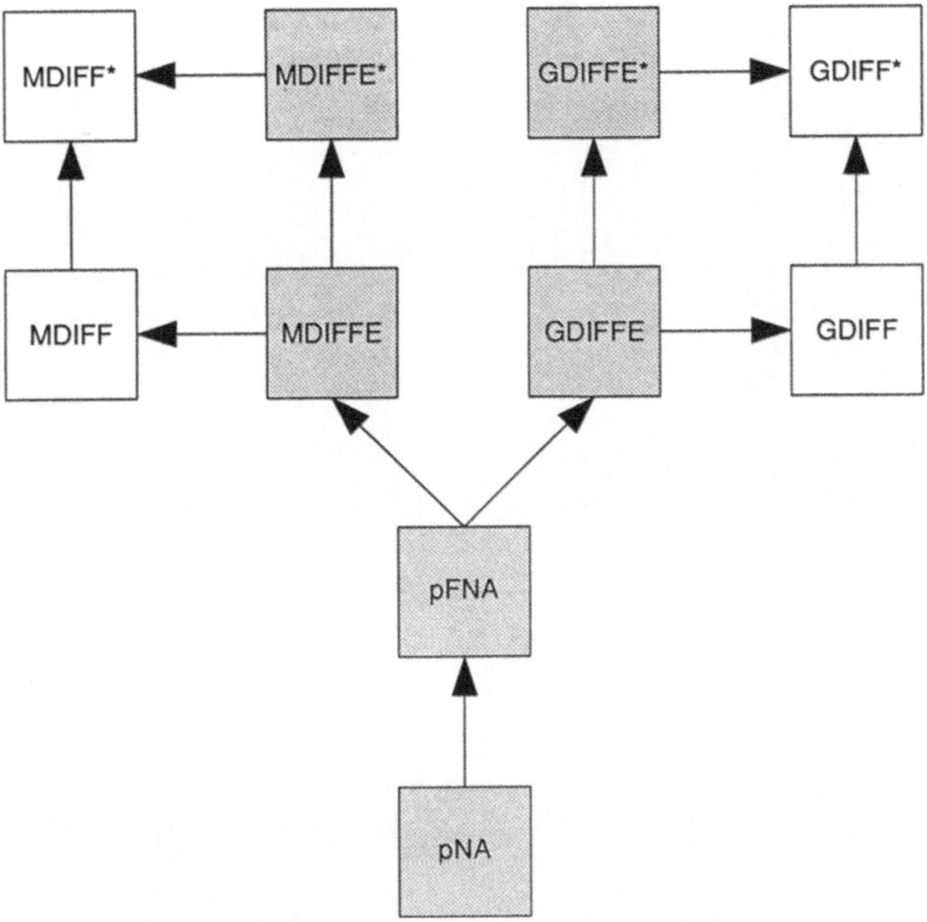

Figure 25.1:
Domains of diagonal Aumann–Shapley values and semivalues (each arrow means a strict inclusion; in spaces marked by white rectangles only semivalues exist).

Chapter VI

Related Topics and Applications

26 Games with Crisp Coalitions: Extensions of the Diagonal Value on pNA

26.1 In this section we discuss some aspects of the problem of defining Aumann–Shapley values on spaces of games with crisp coalitions. By *games with crisp coalitions* we mean games defined on a σ-algebra $\mathscr{A}$ of crisp subsets of X (which is a particular T_∞-tribe). As usual, we write BV and pNA for the spaces $FBV_{\mathscr{A}}$ and $pFNA_{\mathscr{A}}$, respectively. Also, putting $\mathscr{T} = \mathscr{A}^\wedge$, we write FBV and $pFNA$ instead of $FBV_{\mathscr{T}}$ and $pFNA_{\mathscr{T}}$, respectively.

Theorem 18.4 shows that there exists an Aumann–Shapley value ψ on the space pNA of games with crisp coalitions. The space pNA is by no means the largest or the only space of games with crisp coalitions on which an Aumann–Shapley value (even an Aumann–Shapley value having a diagonal representation) exists.

In section 19 we emphasized the very special relationship between the diagonal value φ in $pFNA$ and the Aumann–Shapley value ψ on pNA. To be more specific, we have seen that $\psi = (\varphi \circ e)^\vee$, where $e : pNA \longrightarrow pFNA^*$, is the extension operator introduced in 19.1. This special relationship between ψ and φ was not used when proving the existence of the Aumann–Shapley value ψ on pNA since Theorem 18.4 (from which the existence of ψ follows) does not involve any requirement (concerning the T_∞-tribe) on which the games are defined. When we extended the Aumann–Shapley value φ from $pFNA$ to larger subspaces of FBV, the requirement that the underlying T_∞-tribe be generated was essential (see Theorems 23.6 and 25.5). This means that the extension procedures discussed in Chapter V are not directly applicable to games with crisp coalitions.

However, the special relationship between ψ and φ suggests that a possible way of expanding the Aumann–Shapley value ψ to larger subspaces of BV could be to further extend the extension operator e such that its essential properties (linearity, symmetry, efficiency and positivity) are preserved and the range of this extension is kept within either one of the spaces $GDIFFE^*$ or $MDIFFE^*$. Note that whenever h is a linear, positive, symmetric and efficient operator from a symmetric subspace Q of BV into $GDIFF^*_{\mathcal{T}}$ or $MDIFF^*_{\mathcal{T}}$, then the mappings

$$\boldsymbol{v} \longmapsto [(\theta^* \circ h)(\boldsymbol{v})]^\vee$$

and

$$\boldsymbol{v} \longmapsto [(\xi^* \circ h)(\boldsymbol{v})]^\vee$$

are Aumann–Shapley values on Q, having a diagonal representation (which implies the so-called diagonal property).

26.2 This method of expanding the Aumann–Shapley value ψ on pNA was used in Proposition 44.22 of [AUMANN & SHAPLEY 1974] and later on by MERTENS [1980], showing that there are diagonal values on $pNA \cap pNAD$ and $DIFF$ (and implicitly on $bv'NA$).

The same method is also crucial in MERTENS' [1988] further extension of the Aumann–Shapley value on $DIFF$ to much larger spaces of games with crisp coalitions (which are not necessarily subspaces of BV). The various ways in which the extended extension operator h was defined in these constructions suggests that whenever an Aumann–Shapley value on a space of games with fuzzy coalitions exists, it can be transformed into a tool for obtaining Aumann–Shapley values on suitable spaces of games with crisp coalitions.

In order to demonstrate the power of this method, we use this technique in combination with Theorem 25.5 to build an Aumann–Shapley value on a space we shall call $DIFF'$ which clearly contains both pNA and $pNA' \cap pNAD$ as proper subspaces.

26.3 Recall that pNA' is the closure of pNA in $GAMES_{\mathcal{A}}$ with respect to the topology induced by the *supremum norm* $\|\cdot\|'$ defined by

$$\|\boldsymbol{v}\|' = \sup\{|\boldsymbol{v}(A)| \mid A \in \mathcal{A}\}. \tag{26.1}$$

Since this supremum norm is weaker than the variation norm $\|\cdot\|$, pNA is a proper subspace of pNA'. The extension operator $e : pNA \longrightarrow pFNA$ is still continuous with respect to $\|\cdot\|'$ and, therefore, has a linear continuous extension (also denoted e) to pNA' which happens to preserve properties (19.1), (19.2) as well as positivity, symmetry and efficiency. Let h be the restriction of e to $pNA' \cap BV$. From [AUMANN & SHAPLEY 1974, Lemma 44.2 and Proposition 44.22] it follows that h maps $pNA' \cap BV$ into FBV and preserves the variation norm, and that the linear closed space

$$DIFF' = h^{-1}(GDIFFE_{\mathcal{T}}{}^*) \tag{26.2}$$

contains $pNA' \cap pNAD$ and, therefore, pNA as subspaces.

Proposition. *There exists a unique continuous Aumann–Shapley value ψ' on DIFF′ which satisfies the following diagonal formula:*

$$(\psi' \boldsymbol{v})(A) = \int_0^1 \gamma(h\boldsymbol{v})(t, A) \, \mathrm{d}\lambda(t). \tag{26.3}$$

PROOF: Only the positivity of ψ' is not trivial, and this follows from

$$\psi' \boldsymbol{v} = [(\theta^* \circ h)\boldsymbol{v}]^{\vee},$$

which implies $\|\psi'\| \leq 1$, and Proposition 4.6 in [AUMANN & SHAPLEY 1974]. □

26.4 In spite of repeated successful applications as mentioned above, the method of building Aumann–Shapley values on spaces of games with crisp coalitions by combining suitable extension operators with Aumann–Shapley values obtained on spaces of games with fuzzy coalitions defined on generated T_∞-tribes has its limitations. The fact that there exists an Aumann–Shapley value φ on $pFNA_{\mathcal{T}}$ (as pointed out by BUTNARIU [1987]) which is quite similar to the classical value ψ on pNA, and where no nonatomicity for the underlying σ-additive measure on $\mathcal{A}$ is required (see Theorem 18.4), gave rise to some hope that the method above might allow an Aumann–Shapley value to be built on the closed subspace $pCAM_{\mathcal{A}}$ of BV spanned by the natural powers of not necessarily nonatomic probability measures on $\mathcal{A}$. Such a value would have settled an interesting, still open problem. The hope was mainly based on the fact that the mapping $\bar{e}$ from the set of all natural powers of probability measures on $\mathcal{A}$ to $GDIFFE_{\mathcal{T}}$ given by

$$\bar{e}(\boldsymbol{m}^k)(A) = \left[\int_X A(x) \, \mathrm{d}\boldsymbol{m}(x) \right]^k$$

is well-defined. It was implicitly assumed that $\bar{e}$ has a linear, even continuous, extension to $pCAM_{\mathcal{A}}$. The following example, due to A. WAKOLBINGER [private communication], shows that this is false.

Example. Consider the σ-algebra $\mathcal{A}$ of all crisp subsets of $\{0,1\}$. Define the probability measures $\boldsymbol{m}$, $\boldsymbol{n}$, $\boldsymbol{p}$, $\boldsymbol{q}$ on $\mathcal{A}$ by

$$\begin{aligned} \boldsymbol{m}(\{0\}) &= 0.2, \\ \boldsymbol{n}(\{0\}) &= 0.8, \\ \boldsymbol{p}(\{0\}) &= 0.3, \\ \boldsymbol{q}(\{0\}) &= 0.4. \end{aligned}$$

Then we have

$$\boldsymbol{m}^2 - \boldsymbol{n}^2 - 6\boldsymbol{p} + 6\boldsymbol{q} = \boldsymbol{0},$$

but for the fuzzy coalition $A \in \mathscr{A}^\wedge$ given by $A(0) = 0.5$ and $A(1) = 0$ we obtain

$$\left[\int_X A\,\mathrm{d}\boldsymbol{m}\right]^2 - \left[\int_X A\,\mathrm{d}\boldsymbol{n}\right]^2 - 6\int_X A\,\mathrm{d}\boldsymbol{p} + 6\int_X A\,\mathrm{d}\boldsymbol{q} = 0.05 \neq 0,$$

showing that there is no well-defined linear extension of $\bar{e}$.

27 Values, Cores, and Economic Equilibria

27.1 Since SHAPLEY's [1953] original paper on values for games with finitely many players, the strong relationship between the nature of the *value* of games and their *balancedness* has been well-known. The full meaning of this relationship is revealed in the theory of *transferable utility competitive equilibria* as it evolved from the work of AUMANN & PERLES [1965] and AUMANN & SHAPLEY [1974]. In this section we show that the interrelation between Aumann–Shapley values and balancedness is not specific to games with crisp coalitions. Indeed, it holds for large classes of games with fuzzy coalitions too. We use this fact to give a partial solution to Open Problem C in [AUMANN & SHAPLEY 1974, §41] concerning the balancedness of well-defined market games. In order to provide a proper setup, let us recall the following notions:

Definition. Let $\mathscr{U}$ be a T_∞-tribe (generated or not) on X and assume that $\boldsymbol{v} \in GAMES_{\mathscr{U}}$.

(i) The game $\boldsymbol{v}$ is called *superadditive* if for each pair (A, B) of T_∞-disjoint fuzzy sets in $\mathscr{U}$ we have

$$\boldsymbol{v}(A\,\mathbf{S}_\infty\, B) \geq \boldsymbol{v}(A) + \boldsymbol{v}(B).$$

(ii) The game $\boldsymbol{v}$ is called *homogeneous* (*of degree 1*) if for each $A \in \mathscr{U}$ and for each $\alpha \in [0,1]$ such that $\alpha A \in \mathscr{U}$ we have

$$\boldsymbol{v}(\alpha A) = \alpha \boldsymbol{v}(A).$$

(iii) The game $\boldsymbol{v}$ is called *uniformly continuous* if $\boldsymbol{v}$ is continuous with respect to the *uniform topology* induced on $\mathscr{U}$ by the supremum norm

$$\|A\| = \sup\{A(x) \mid x \in X\}.$$

(iv) The game $\boldsymbol{v}$ is called *balanced* if there exists a finitely T_∞-additive function $\boldsymbol{m} : \mathscr{U} \longrightarrow \mathbb{R}$ such that

$$\boldsymbol{m}(X) \leq \boldsymbol{v}(X) \tag{27.1}$$

and for all $A \in \mathscr{U}$

$$\boldsymbol{m}(A) \geq \boldsymbol{v}(A). \tag{27.2}$$

The (possibly empty) set of all finitely T_∞-functions $\boldsymbol{m} : \mathscr{U} \longrightarrow \mathbb{R}$ satisfying (27.1) and (27.2) is called the *core of* $\boldsymbol{v}$ and is denoted Core($\boldsymbol{v}$).

(v) The game $\boldsymbol{v}$ is called *σ-balanced* if Core($\boldsymbol{v}$) contains a finite T_∞-measure on $\mathscr{U}$. The (possibly empty) set of T_∞-measures in Core($\boldsymbol{v}$) is denoted $\text{Core}_\sigma(\boldsymbol{v})$ and is called the *σ-core of* $\boldsymbol{v}$.

27.2 To decide whether a game is balanced or not is not at all trivial, even in the case of games with finitely many players and crisp coalitions only (see [BANDAREVA 1963], [SHAPLEY 1971, 1973], [BUTNARIU & CENSOR 1991]). A traditional way of doing this consists of showing the existence of an Aumann–Shapley value for a given game $\boldsymbol{v}$ which belongs to Core($\boldsymbol{v}$) (see [SHAPLEY 1953], [AUMANN & SHAPLEY 1974], [MERTENS 1988]). An alternative approach is usually based on the study of specific analytical properties of various classes of games (see [BONDAREVA 1963], [PARKER 1991]). It is even more difficult to decide whether a game is σ-balanced. KANNAI [1966] and SCHMEIDLER [1969, 1972] have shown that σ-balanced games are quite special, and they proved criteria for the σ-balancedness of games with crisp coalitions which can be carried over to games with fuzzy coalitions without essential difficulties.

27.3 The following result shows that a similar relationship between values and balancedness, as emphasized by SHAPLEY [1953] in the case of games with finitely many players and by AUMANN & SHAPLEY [1974, Theorem I] in the case of games in *pNA*, still holds for Gâteaux differentiable games.

Theorem. *Let $\mathscr{T}$ be a generated T_∞-tribe and let $\boldsymbol{v} \in GDIFFE_{\mathscr{T}}$. If $\boldsymbol{v}$ is superadditive and homogeneous, then we have:*

(i) the game $\boldsymbol{v}$ is balanced and the diagonal value $\theta\boldsymbol{v}$ of $\boldsymbol{v}$ belongs to Core($\boldsymbol{v}$);

(ii) $\theta\boldsymbol{v}$ is the unique uniformly continuous finitely T_∞-additive function which belongs to Core($\boldsymbol{v}$);

(iii) if $\boldsymbol{v}$ is σ-balanced, then $\text{Core}_\sigma(\boldsymbol{v}) = \{\theta\boldsymbol{v}\}$;

(iv) if $\boldsymbol{v}$ is uniformly continuous, then Core($\boldsymbol{v}$) $= \{\theta\boldsymbol{v}\}$.

PROOF: The proof will be organized in several steps.

Claim 1. If $\boldsymbol{v} \in GDIFFE_{\mathcal{T}}$ is homogeneous and if for $t \in [0,1]$ the Dini derivative $\delta\boldsymbol{v}(t,\cdot)$ exists, then

$$\theta\boldsymbol{v}(\cdot) = \delta\boldsymbol{v}(t,\cdot). \tag{27.3}$$

In order to verify this, recall that Proposition 23.2 ensures that $\delta\boldsymbol{v}(t,\cdot)$ exists for λ-almost all $t \in]0,1[$ whenever $\boldsymbol{v} \in GDIFFE_{\mathcal{T}}$. Let $t_1, t_2 \in]0,1]$ with $t_1 < t_2$ such that both $\delta\boldsymbol{v}(t_1,\cdot)$ and $\delta\boldsymbol{v}(t_2,\cdot)$ exist, then the homogeneity of $\boldsymbol{v}$ yields for $\tau \in]0, 1-t_2[$ and $A \in \mathcal{T}$

$$\begin{aligned}
\frac{\boldsymbol{v}(t_1X\,\mathbf{S}_\infty\,\tau A) - \boldsymbol{v}(t_1A)}{\tau} &= \frac{\boldsymbol{v}(t_1X + \tau A) - \boldsymbol{v}(t_1X)}{\tau} \\
&= \frac{\frac{t_1}{t_2}\cdot\boldsymbol{v}\left[t_2X + \left(\frac{t_2}{t_1}\cdot\tau\right)\cdot A\right] - \frac{t_1}{t_2}\cdot\boldsymbol{v}(t_2X)}{\tau} \\
&= \frac{\boldsymbol{v}\left[t_2X\,\mathbf{S}_\infty\left(\frac{t_2}{t_1}\cdot\tau\right)\cdot A\right] - \boldsymbol{v}(t_2A)}{\frac{t_2}{t_1}\cdot\tau}.
\end{aligned}$$

Taking the limit $\tau \searrow 0$ and taking into account the existence of $\delta\boldsymbol{v}(t_1,A)$ and $\delta\boldsymbol{v}(t_2,A)$, it follows that

$$\delta\boldsymbol{v}(t_1,A) = \delta\boldsymbol{v}(t_2,A),$$

showing that for each $A \in \mathcal{T}$, $\delta\boldsymbol{v}(\cdot,A)$ is constant λ-a.e. (to be more specific, $\delta\boldsymbol{v}(\cdot,A)$ is constant wherever it exists). This, together with (23.6) yields (27.3) for each $t \in [0,1]$ where $\delta\boldsymbol{v}(t,\cdot)$ exists, thus completing the proof of Claim 1.

From Proposition 23.2 we know that for $\boldsymbol{v} \in GDIFF_{\mathcal{T}}$ the Dini derivative $\delta\boldsymbol{v}(t,\cdot)$ exists for λ-almost all $t \in [0,1]$. Then Claim 1 implies that

$$\theta\boldsymbol{v}(\cdot) = \delta\boldsymbol{v}(t,\cdot) \qquad \text{for } \lambda\text{-almost all } t \in [0,1]. \tag{27.4}$$

We are now ready to prove (i). Note that $\theta\boldsymbol{v}$ is a finite T_∞-additive function on $\mathcal{T}$ which satisfies (27.1) because of the efficiency of θ. In order to show that $\theta\boldsymbol{v}$ also satisfies (27.2), choose $A \in \mathcal{T}$, $t \in]0,1[$ such that $\delta\boldsymbol{v}(t,\cdot)$ exists, and $\tau \in]0,1-t[$. Then tX and τA are T_∞-disjoint, and we have

$$\begin{aligned}
\boldsymbol{v}(A) &= \frac{\boldsymbol{v}(tX) + \tau\boldsymbol{v}(A) - \boldsymbol{v}(tX)}{\tau} \\
&= \frac{\boldsymbol{v}(tX) + \boldsymbol{v}(\tau X) - \boldsymbol{v}(tX)}{\tau} \\
&\leq \frac{\boldsymbol{v}(tX\,\mathbf{S}_\infty\,\tau A) - \boldsymbol{v}(tX)}{\tau}
\end{aligned}$$

since $\boldsymbol{v}$ is homogeneous and superadditive. Taking the limit $\tau \searrow 0$ and observing that $\delta\boldsymbol{v}(t,\cdot)$ exists, we obtain

$$\boldsymbol{v}(A) \leq \delta\boldsymbol{v}(t,A).$$

Hence, according to (27.4), it follows that $\boldsymbol{\theta v}$ satisfies (27.2), i.e., $\boldsymbol{\theta v} \in \text{Core}(\boldsymbol{v})$ and the validity of (i) is shown.

The proof of (ii) and (iii) will be based on the following:

Claim 2. If $\boldsymbol{m} : \mathscr{T} \longrightarrow \mathbb{R}$ is a finitely T_∞-additive function, then for each $\alpha \in [0,1] \cap \mathbb{Q}$ and for all $A \in \mathscr{T}$ we have

$$\boldsymbol{m}(\alpha A) = \alpha \cdot \boldsymbol{m}(A). \tag{27.5}$$

Moreover, if $\boldsymbol{m}$ is also uniformly continuous, then $\boldsymbol{m}$ is homogeneous.

For all $A \in \mathscr{T}$ and $k, l \in \mathbb{N}$ with $k \leq l$ the family of fuzzy coalitions $(A_1, A_2, \ldots, A_k)$ defined by $A_i = \frac{1}{l} \cdot A$ for each $i \in \{1, 2, \ldots, k\}$ is T_∞-disjoint and we have

$$\boldsymbol{m}\left(\frac{k}{l} \cdot A\right) = \boldsymbol{m}(\mathbf{S}_{\infty \atop i=1}^{l} A_i) = \sum_{i=1}^{l} \boldsymbol{m}(A_i) = k \cdot \boldsymbol{m}\left(\frac{1}{l} \cdot A\right). \tag{27.6}$$

In particular, for $k = l$ this gives

$$\frac{1}{l} \cdot \boldsymbol{m}(A) = \boldsymbol{m}\left(\frac{1}{l} \cdot A\right),$$

which together with (27.6) implies (27.5). Now, suppose that $\boldsymbol{m}$ is also uniformly continuous. For each $\alpha \in [0,1]$ there exists a sequence $(\alpha_n)_{n \in \mathbb{N}} \in [0,1] \cap \mathbb{Q}$ converging increasingly to α. Since for $A \in \mathscr{T}$ the sequence $(\alpha_n \cdot A)_{n \in \mathbb{N}}$ of fuzzy coalitions converges uniformly to the fuzzy coalition $\alpha \cdot A$, we get

$$\boldsymbol{m}(\alpha \cdot A) = \lim_{n \to \infty} \boldsymbol{m}(\alpha_n \cdot A) = \lim_{n \to \infty} \alpha_n \cdot \boldsymbol{m}(A) = \alpha \cdot \boldsymbol{m}(A),$$

where the first equality follows from the uniform continuity and the second one from (27.5). This completely proves Claim 2.

Now we use Claim 2 to show (ii). Let $\boldsymbol{m} \in \text{Core}(\boldsymbol{v})$ be uniformly continuous. Choose $t \in]0,1[$ such that $\delta \boldsymbol{v}(t, \cdot)$ exists and $\tau \in]0, 1-t[$. Then $\boldsymbol{m}$ and $\boldsymbol{v}$ are both homogeneous and we have:

$$t \cdot \boldsymbol{v}(X) = \boldsymbol{v}(tX) = \boldsymbol{m}(tX) = t \cdot \boldsymbol{m}(X), \tag{27.7}$$

since from (27.2) and (27.1) we can deduce

$$t \cdot \boldsymbol{m}(X) = \boldsymbol{m}(tX) \geq \boldsymbol{v}(tX) = t \cdot \boldsymbol{v}(X) = t \cdot \boldsymbol{m}(X).$$

Also, for each $A \in \mathscr{T}$, we get

$$\begin{aligned}
\boldsymbol{m}(A) &= \frac{\boldsymbol{m}(tX) + \tau \cdot \boldsymbol{m}(A) - \boldsymbol{m}(tX)}{\tau} \\
&= \frac{\boldsymbol{m}(tX) + \boldsymbol{m}(\tau A) - \boldsymbol{m}(tX)}{\tau} \\
&= \frac{\boldsymbol{m}(tX \, \mathbf{S}_\infty \, \tau A) - \boldsymbol{m}(tX)}{\tau}
\end{aligned}$$

since tX and τA are T_∞-disjoint. Together with (27.2) and (27.7) this implies

$$\boldsymbol{m}(A) \geq \frac{\boldsymbol{v}(tX\,\mathbf{S}_\infty\,\tau A) - \boldsymbol{v}(tX)}{\tau}$$

and, taking the limit $\tau \searrow 0$ and using (27.3), we obtain for each $A \in \mathscr{T}$

$$\boldsymbol{m}(A) \geq \delta\boldsymbol{v}(t, A) = \theta\boldsymbol{v}(A). \tag{27.8}$$

Since (27.8) holds for each $A \in \mathscr{T}$, we also have $\boldsymbol{m}(A^c) \geq \theta\boldsymbol{v}(A^c)$ and this, together with (27.8), implies $\boldsymbol{m}(A) = \theta\boldsymbol{v}(A)$ for all $A \in \mathscr{T}$. Hence, if $\boldsymbol{m}$ is uniformly continuous, then $\boldsymbol{m} = \theta\boldsymbol{v}$. Obviously, Proposition 23.2 and Theorem 23.6 imply that $\theta\boldsymbol{v}$ is uniformly continuous, and (ii) is proven. Observe that each $\boldsymbol{m} \in \mathrm{Core}_\sigma(\boldsymbol{v})$ can be represented by Corollary 10.3, showing that $\boldsymbol{m}$ is uniformly continuous. Hence, for $\boldsymbol{m} \in \mathrm{Core}_\sigma(\boldsymbol{v})$ we have $\boldsymbol{m} = \theta\boldsymbol{v}$, and this proves (iii).

It remains to show (iv). Suppose that $\boldsymbol{v}$ is uniformly continuous and that $\boldsymbol{m} \in \mathrm{Core}(\boldsymbol{v})$. From (i) we know that $\theta\boldsymbol{v} \in \mathrm{Core}(\boldsymbol{v})$. We want to show that $\boldsymbol{m} = \theta\boldsymbol{v}$. For this purpose, assume that $t \in]0,1[$ and $\delta\boldsymbol{v}(t,\cdot)$ exists. Choose $\tau \in]0, 1-t[\cap\mathbb{Q}$ and let $(t_n)_{n\in\mathbb{N}}$ be a sequence in $]0,1[\cap\mathbb{Q}$ which converges increasingly to t. Then, for each $n \in \mathbb{N}$, $\tau \in]0, 1-t_n[$ and Claim 2 yields

$$\boldsymbol{m}(t_nX) = t_n \cdot \boldsymbol{m}(X).$$

Hence, for each $A \in \mathscr{T}$ and for each $n \in \mathbb{N}$ we have

$$\begin{aligned} \boldsymbol{m}(A) &= \frac{\boldsymbol{m}(t_nX\,\mathbf{S}_\infty\,\tau A) - \boldsymbol{m}(t_nX)}{\tau} \\ &\geq \frac{\boldsymbol{v}(t_nX\,\mathbf{S}_\infty\,\tau A) - \boldsymbol{v}(t_nX)}{\tau}. \end{aligned}$$

Taking the limit $n \to \infty$ we obtain for each $A \in \mathscr{T}$

$$\boldsymbol{m}(A) \geq \frac{\boldsymbol{v}(tX\,\mathbf{S}_\infty\,\tau A) - \boldsymbol{v}(tX)}{\tau}$$

since $\boldsymbol{v}$ is uniformly continuous. Taking the limit $\tau \searrow 0$ in the last inequality, we get

$$\boldsymbol{m}(A) \geq \delta\boldsymbol{v}(t, A)$$

for all $A \in \mathscr{T}$, and this implies $\boldsymbol{m} = \theta\boldsymbol{v}$ because of Claim 1. □

27.4 Remark.

(i) Theorem 27.3 is proved for games on generated T_∞-tribes only. However, if $\boldsymbol{v}$ is a game on a nongenerated T_∞-tribe $\mathscr{T}$ and if $\boldsymbol{v}$ has an "extension" $\boldsymbol{v}^*$ to $\mathscr{T}^\circ$ such that $\boldsymbol{v}^* \in GDIFFE_{\mathscr{T}^\circ}$ and $\boldsymbol{v}^*$ is superadditive and homogeneous, then $\boldsymbol{v}$ is balanced and $(\theta\boldsymbol{v}^*)^\vee$, the restriction of $\theta\boldsymbol{v}^*$ to $\mathscr{T}$, belongs to $\mathrm{Core}(\boldsymbol{v})$.

(ii) Theorem 27.3 (iv) applies in particular to games in $pFNA_{\mathcal{T}}$ which are uniformly continuous and belong to $GDIFFE_{\mathcal{T}}$. To be precise, if $\boldsymbol{v} \in pFNA_{\mathcal{T}}$ is superadditive and homogeneous, then $\text{Core}(\boldsymbol{v}) = \{\varphi\boldsymbol{v}\}$, where φ is the diagonal value on $pFNA_{\mathcal{T}}$. This suggests an alternative proof of Theorem·I of AUMANN & SHAPLEY [1974]. Note that games $\boldsymbol{v} \in pNA$ have (uniformly continuous) extensions

$$\boldsymbol{v}^* = e(\boldsymbol{v}) \in pFNA^*_{\mathcal{T}} \subseteq pFNA_{\mathcal{T}} \subseteq GDIFFE_{\mathcal{T}}$$

(see Section 19). If $\boldsymbol{v}$ is superadditive, then so is $\boldsymbol{v}^*$ (see [AUMANN & SHAPLEY 1974, Proposition 27.1]). If $\boldsymbol{v}^*$ is also homogeneous, then $\text{Core}_\sigma(\boldsymbol{v}^*) = \{\varphi\boldsymbol{v}^*\}$ and $(\varphi\boldsymbol{v}^*)^\vee \in \text{Core}_\sigma(\boldsymbol{v})$. In order to show that $(\varphi\boldsymbol{v}^*)^\vee$ is the only element of $\text{Core}(\boldsymbol{v})$, it suffices to prove that for a superadditive $\boldsymbol{v} \in pNA_{\mathcal{T}}$ and $\boldsymbol{m} \in \text{Core}(\boldsymbol{v})$ we have $e(\boldsymbol{m}) \in \text{Core}(\boldsymbol{v}^*)$. But this is a consequence of the *NA*-continuity of $\boldsymbol{v}^*$ (see [AUMANN & SHAPLEY 1974, §27]).

(iii) SHAPLEY [1953] has shown that if $\mathcal{T}$ is the σ-algebra of all crisp subsets of X and if $\boldsymbol{v} \in GAMES_{\mathcal{T}}$ is superadditive, then $\chi\boldsymbol{v} \in \text{Core}(\boldsymbol{v})$ (see (19.9)). This result can be obtained from Theorem 27.3, observing that whenever $\boldsymbol{v} \in GAMES_{\mathcal{T}}$ is superadditive, then the multilinear extension of $\boldsymbol{v}$ is superadditive and homogeneous and belongs to $GDIFFE_{\mathcal{T}^\wedge}$. Therefore, (i) above applies.

27.5 AUMANN & SHAPLEY [1974, Proposition 32.5] established a strong interconnection between the existence of *transferable utility competitive equilibria* in *transferable utility economies*, on the one hand, and the balancedness of certain classes of games, on the other hand.

Recall that a (*transferable utility*) *economy* (with n *types of commodities*) is described by a measurable space $(X, \mathcal{A}, \boldsymbol{q})$, a *utility function* $u : X \times]0,\infty[^n \longrightarrow [0,\infty[$ and an *initial allocation* $a : X \longrightarrow [0,\infty[^n$. In this setting the set X is the set of *economic agents* (*players*), the σ-algebra $\mathcal{A}$ of crisp subsets of X is the family of *feasible coalitions* and the finite σ-additive measure $\boldsymbol{q}$ is called *population measure*. The utility function u is presumed to be ($\mathcal{A} \times Borel$)–measurable, for each economic agent $x \in X$ the function $u(x,\cdot)$ is assumed to be continuous and nondecreasing, and the initial allocation a has to be $\boldsymbol{q}$-integrable with

$$\int_X a\,\mathrm{d}\boldsymbol{q} > 0.$$

All integrals of vector–valued functions from now on are thought of as being taken component by component, the value of the integral, therefore, being itself a vector. Strict inequalities between vectors are understood to hold for each component.

An *allocation* in the given economy is a $\boldsymbol{q}$-integrable function $f : X \longrightarrow [0,\infty[^n$ such that

$$\int_X f\,\mathrm{d}\boldsymbol{q} = \int_X a\,\mathrm{d}\boldsymbol{q}. \tag{27.9}$$

A *transferable utility competitive equilibrium* (t.u.c.e.) is a pair (f,p) where f is an allocation and $p \in]0,\infty[^n$ is a *price vector*, such that for each $x \in X$ we have

$$u(x,f(x)) - p(f(x)-a(x)) = \max\{u(x,z) - p\cdot(z-a(x)) \mid z \in]0,\infty[^n\}, \quad (27.10)$$

where $p\cdot(z-a(x))$ is the scalar product. If (f,p) is a transferable utility competitive equilibrium, then the σ-additive measure $\boldsymbol{m} : A \longrightarrow \mathbb{R}$ defined by

$$\boldsymbol{m}(A) = \int_A [u(x,f(x)) - p(f(x)-a(x))]\,\mathrm{d}\boldsymbol{q}(x) \quad (27.11)$$

is called the *competitive payoff distribution* of (f,p).

Denote $\mathscr{T} = \mathscr{A}^\wedge$ and for each $A \in \mathscr{T}$ let $\mathscr{F}(A)$ be the set of all *feasible allocations for the coalition* A, i.e., the set of all allocations $f : X \longrightarrow]0,\infty[^n$ such that

$$\int_X A\cdot f\,\mathrm{d}\boldsymbol{q} = \int_X A\cdot a\,\mathrm{d}\boldsymbol{q}. \quad (27.12)$$

The *extended market game* associated with a given economy is the function $\hat{\boldsymbol{v}} : \mathscr{T} \longrightarrow [0,\infty]$ defined by

$$\hat{\boldsymbol{v}}(A) = \sup\left\{\int_X A(x)\cdot u(x,f(x))\,\mathrm{d}\boldsymbol{q}(x) \mid f \in \mathscr{F}(A)\right\}. \quad (27.13)$$

The restriction $\boldsymbol{v} = (\hat{\boldsymbol{v}})^\vee$ of $\hat{\boldsymbol{v}}$ to $\mathscr{A}$ is usually called the *market game* (associated with the given economy). Note that either $\boldsymbol{v}$ or $\hat{\boldsymbol{v}}$ may assume infinite values too. The extended market game is said to be *well-defined* if for each $A \in \mathscr{T}$ there exists a feasible allocation $f \in \mathscr{F}(A)$ at which the supremum in (27.13) is attained and finite, i.e.,

$$\hat{\boldsymbol{v}}(A) = \int_X A(x)\cdot u(x,f(x))\,\mathrm{d}\boldsymbol{q}(x). \quad (27.14)$$

If, given a market game $\boldsymbol{v}$, for each $A \in \mathscr{A}$ there is an $f \in \mathscr{F}(A)$ such that for the extened market game $\hat{\boldsymbol{v}}$ equation (27.14) holds, we say that $\boldsymbol{v}$ is *well-defined*.

27.6 AUMANN & SHAPLEY [1974, Proposition 32.5 and Theorem J] have shown that, in some cases, for economies with nonatomic population measure $\boldsymbol{q}$ and well-defined associated market game $\boldsymbol{v}$, Core($\boldsymbol{v}$) is exactly the set of all competitive payoff distributions of all transferable utility competitive equilibriums in the given economy. Moreover, if for each $x \in X$ its utility function $u(x,\cdot)$ has continuous partial derivatives $\frac{\partial u}{\partial z_j}(x,\cdot)$ for all $j \in \{1,2,\ldots,n\}$ and if $u(x,z) = o(\|z\|)$ as $\|z\| \to \infty$ (integrably in x), then $\boldsymbol{v} \in pNA$ and, consequently, $\varphi\boldsymbol{v} \in$ Core($\boldsymbol{v}$) is the competitive payoff distribution of a t.u.c.e. in the given economy. The strong restrictions imposed upon the population measure $\boldsymbol{q}$ and the utility function u of the given economy are mostly technical, i.e., they are sufficient conditions to ensure that $\boldsymbol{v}$ is well-defined and belongs to *pNA*. AUMANN & SHAPLEY [1974, §41, Open Problem C] themselves raised the question of whether a result similar to their Theorem J could be obtained under weaker hypotheses, especially by removing the ***asymptotic condition*** that $u(x, z) = o(\|z\|)$ as $\|z\| \to \infty$, integrably in x, and replacing it by the condition that the market game $\boldsymbol{v}$ be well-defined. The asymptotic condition, together with the hypothesis, implies that the extended market game $\hat{\boldsymbol{v}}$ and, a fortiori, the market game $\boldsymbol{v}$ are well-defined.

27.7 In what follows we give a partial solution of the open problem mentioned above showing that, when replacing the asymptotic condition by the weaker well-definedness of the ***extended market game*** $\hat{\boldsymbol{v}}$ (and even weakening some of the other hypotheses), the implications of Theorem J of AUMANN & SHAPLEY [1974] still hold. Our solution is only partial in the sense that we do not know whether, in general, the well-definedness of $\boldsymbol{v}$ implies the well-definedness of $\hat{\boldsymbol{v}}$. To be precise, we have the following result which requires neither the standardness assumption nor that the population measure $\boldsymbol{q}$ be nonatomic.

Proposition. *Assume that for each $x \in X$, the partial derivatives $\frac{\partial u}{\partial z_j}(x, \cdot)$ of its utility function $u(x, \cdot)$ exist and are finite on $]0, \infty[^n$ for all $j \in \{1, 2, \ldots, n\}$. If the extended market game $\hat{\boldsymbol{v}}$ is well-defined, then $\hat{\boldsymbol{v}} \in GDIFFE_{\mathcal{T}}$ and its diagonal value $\theta\hat{\boldsymbol{v}} \in Core(\hat{\boldsymbol{v}})$. Moreover, the restriction $(\theta\hat{\boldsymbol{v}})^{\vee}$ of $\theta\hat{\boldsymbol{v}}$ to $\mathcal{A}$ equals the competitive payoff distribution of a t.u.c.e. in the given economy.*

PROOF: It follows from inequality (32.16) in [AUMANN & SHAPLEY 1974] which is, in fact, an equality, that $\hat{\boldsymbol{v}} \in GDIFFE_{\mathcal{T}}$. Obviously, $\hat{\boldsymbol{v}}$ is superadditive and homogeneous. Hence, Theorem 27.3 (i) shows that $\theta\hat{\boldsymbol{v}} \in \text{Core}(\hat{\boldsymbol{v}})$. As a consequence, the restriction $\boldsymbol{v}$ of $\hat{\boldsymbol{v}}$ to $\mathcal{A}$ is well-defined too, and $(\theta\hat{\boldsymbol{v}})^{\vee} \in \text{Core}(\boldsymbol{v})$. By Proposition 32.5 in [AUMANN & SHAPLEY 1974], which still holds in our setting, Core($\boldsymbol{v}$) equals the set of all competitive payoff distributions of a t.u.c.e. in the given economy. □

28 Rate Problem for Services in Bulk: A Game Theoretical Model

28.1 BILLERA, HEATH & RAANAN [1978] applied Aumann–Shapley values to solve the problem of internal telephone billing rates, which is a special case of the general rate problem for services in bulk. This general rate problem has been formulated by them as follows:

> *There is a group of customers or consumers requiring some service. The service cannot be provided to each customer separately in exactly the*

amount he needs, but rather is available in bulk. By that we mean that it can either be supplied in big 'chunks' or that there is a very high initial fee involved that makes it unprofitable for any single customer to purchase that service by himself. However, if the customers form a coalition, their total demand may be high enough to make acquiring that service feasible. The question that arises now is how to charge each customer his 'fair share' of cost, given that we have to recover all of it from users' payments.

Examples of rate problems for services in bulk are the determination of taxes for municipial services, the fixing of prices for utilities (energy, water, etc.) in industrial organizations and internal telephone billing rates as considered by BILLERA et al. [1978].

In fact, internal telephone billing rates are a special case since the individual services (telephone calls) are either entirely purchased by a single customer or not, no intermediate situation being possible. A different situation arises when services such as energy or water are considered which are provided continuously and then shared among the users (see [BUTNARIU 1990]).

In contrast to the case of telephone calls, where each individual service is produced for and purchased by a single customer (and, consequently, belongs entirely to this customer), with utilities each amount of energy or water is produced for many consumers who share it according to their individual needs at a given moment. This can be interpreted in such a way that each amount of the utility produced belongs to a certain customer to a certain degree representing his or her share of the given amount of service.

In spite of this difference, the same *rationality criteria* should be applied when computing rates for telephone calls or *continuous* utilities: *efficiency*, i.e., the total cost of providing services (including effective costs, profit, taxes, etc.) has to be equal to the consumers' total payment; *symmetry*, i.e., identical services are charged identically whenever they are consumed; *positivity*, i.e., higher amounts of a given type of service imply higher payoffs; *linearity*, i.e., the rate charged to a customer does not depend on how the total cost is broken down into its components.

28.2 In order to build a game theoretical model for the rate problem of services in bulk, we interpret the individual services as players in a cooperative game, coalitions of which are the bulks of services which are eventually provided. Customers will be identified with the specific bulks of services they consume. Since different customers may sometimes share the same service (as in the case of energy or water supply), we have to deal with fuzzy coalitions representing bulks for services where the degree of membership of an individual service is exactly the share of this service consumed by a given customer. The total bulk of all services produced and provided in a specific interval of time (i.e., the set of players) is denoted X, as usual. Fuzzy coalitions consequently are fuzzy subsets of X, and we presume that all fuzzy coalitions form a suitable generated T_∞-tribe $\mathscr{T}$. The worth $\boldsymbol{v}(A)$ of a fuzzy coalition A in the game $\boldsymbol{v} : \mathscr{T} \longrightarrow \mathbb{R}$ is represented by the total cost involved in producing and providing the bulk of services A. It will turn out that the rate problem of services in bulk can be solved by computing the number $\boldsymbol{v}(A)$ for a very limited number of coalitions only.

28.3 We must take into account two kinds of costs for each service $x \in X$: the direct costs which are related to the production of the specific service x (and, therefore, must be charged to the consumers of x), and the indirect costs for producing the services in X (usually costs for machine equipment, salaries, etc.) which have to be charged to all customers depending (not necessarily proportionally) on the share of services which they consume. Theoretically speaking, this implicitly means that we end up with an individual rate $r(x)$ for each service $x \in X$. From a practical point of view, there are usually similar services with similar costs, which consequently leads to almost identical rates. This suggests partitioning the set of services X into a finite number of *types of services*, say $P_1, P_2, \ldots, P_n$. In what follows, we always assume that such a partition is, a priori, given. Therefore, the problem consists of finding rates $r_1, r_2, \ldots, r_n$ where r_i is the rate for services of the type P_i.

28.4 We assume that for each $i \in \{1, \ldots, n\}$ there is a σ-additive measure $\boldsymbol{m}_i : \mathscr{T}^\vee \longrightarrow \mathbb{R}$ such that $\boldsymbol{m}_i(P_j) = 0$ whenever $j \neq i$, assigning to each crisp bulk (= coalition) $A \in \mathscr{T}^\vee$ the amount of services of type P_i contained in A. Since the sets $P_1, P_2, \ldots, P_n$ are mutually disjoint, the assumption $\boldsymbol{m}_i(P_j) = 0$ for $j \neq i$ is quite natural.

This assumption can be made in many practical cases. For instance, in the discrete case of internal telephone billing rates (see [BILLERA et al. 1978]), a type P_i consists of all phonc calls to a specific destination completed during a given time interval of the day, and $\boldsymbol{m}_i(A)$ represents the total duration of calls of type P_i within the bulk A.

In other problems, such as rating energy or water where the services are provided in continuously varying amounts, the types of services can be defined similarly according to the period of the day (or week or year) and to the transportation costs. In such cases, $\boldsymbol{m}_i(A)$ can be obtained using different metering systems for different types of services, or simply by defining $\boldsymbol{m}_i(A)$ according to the proportion of services of type P_i in the global amount of services produced.

Since we sometimes have to deal with a T_∞-tribe $\mathscr{T}$ of fuzzy coalitions (= bulks), as noted in 28.2, we have to *measure* the amount of services of type P_i in a fuzzy coalition A. To this end, the usual differential approach is useful: if $\mathrm{d}x$ is a suitable infinitesimally small crisp bulk of services containing a service x, then

$$\mathrm{d}\boldsymbol{m}_i(x) = \boldsymbol{m}_i(\mathrm{d}x)$$

represents the amount of services of type P_i in dx and since degrees of membership in A can be assumed to be nearly constant on dx,

$$A(x) \cdot \mathrm{d}\boldsymbol{m}_i(x)$$

approximates the share of the quantity of services of type P_i in dx contained in A. Hence the *summation*

$$\hat{\boldsymbol{m}}_i(A) = \int_X A(x)\,\mathrm{d}\boldsymbol{m}_i(x) \tag{28.1}$$

yields the total amount of services of type P_i *contained* in the fuzzy coalition A. Note that the measures $\boldsymbol{m}_i$ may be atomic while the functions $\hat{\boldsymbol{m}}_i$ are necessarily nonatomic T_∞-measures on $\mathscr{T}$ since $\mathscr{T}$ is generated.

28.5 The toal cost (including both the direct and the indirect cost) of producing and providing the amounts $q_1, q_2, \ldots, q_n$ of the services of type $P_1, P_2, \ldots, P_n$, respectively, will be denoted $f(q_1, q_2, \ldots, q_n)$, and is obtained either analytically or, if this is not possible, by regression analysis.

If A is a fuzzy coalition then necessarily

$$\boldsymbol{v}(A) = f(\hat{\boldsymbol{m}}(A)), \tag{28.2}$$

where $\hat{\boldsymbol{m}} = (\hat{\boldsymbol{m}}_1, \hat{\boldsymbol{m}}_2, \ldots, \hat{\boldsymbol{m}}_n)$ is the n-vector T_∞-measure on $\mathscr{T}$ whose components $\hat{\boldsymbol{m}}_i$ are defined in (28.1).

Whenever the function f has measurable finite partial derivatives $\frac{\partial f}{\partial q_i}$ at each point of the range $\mathcal{R}(\hat{\boldsymbol{m}})$, the game $\boldsymbol{v}$ is an element of the space $GDIFFE_{\mathscr{T}}$ (see Theorem 25.4). In the sequel we assume that this is always the case. This implies the existence of an Aumann–Shapley value $\zeta\boldsymbol{v}$ for the game $\boldsymbol{v}$ with fuzzy coalitions which can be computed by the diagonal formula

$$\zeta\boldsymbol{v}(A) = \sum_{i=1}^{n} \hat{\boldsymbol{m}}_i(A) \cdot \int_0^1 \frac{\partial}{\partial q_i} f(t \cdot \hat{\boldsymbol{m}}(X))\,\mathrm{d}t. \tag{28.3}$$

Being an Aumann–Shapley value, $\zeta\boldsymbol{v}$ satisfies the rationality requirements mentioned in subsection 28.1.

28.6 Now, for each $i \in \{1, 2, \ldots, n\}$ define

$$r_i = \int_0^1 \frac{\partial f}{\partial q_i}(t \cdot \hat{\boldsymbol{m}}(X))\,\mathrm{d}t \tag{28.4}$$

and insert this into (28.3):

$$\zeta\boldsymbol{v}(A) = \sum_{i=1}^{n} \hat{\boldsymbol{m}}_i(A) \cdot r_i. \tag{28.5}$$

The number r_i is exactly the rate for the services of type P_i since for the coalition P_i we get

$$\zeta \boldsymbol{v}(P_i) = \hat{\boldsymbol{m}}_i(P_i) \cdot r_i,$$

and thus

$$r_i = \frac{\zeta \boldsymbol{v}(P_i)}{\hat{\boldsymbol{m}}_i(P_i)}$$

whenever the denominator $\hat{\boldsymbol{m}}_i(P_i) \neq 0$, i.e., r_i represents the total cost divided by the total quantity of services of type P_i.

From a computational point of view, formula (28.4) is quite simple: it involves only the partial derivatives $\frac{\partial f}{\partial q_i}$ of the *cost function* f and the vector

$$\hat{\boldsymbol{m}}(X) = (\hat{\boldsymbol{m}}_1(P_1), \hat{\boldsymbol{m}}_2(P_2), \ldots, \hat{\boldsymbol{m}}_n(P_n)).$$

28.7 If the cost function f is linear, i.e., if

$$f(q_1, q_2, \ldots, q_n) = \sum_{i=1}^{n} \alpha_i \cdot q_i,$$

then it is clear that $r_i = \alpha_i$ for each $i \in \{1, 2, \ldots, n\}$. This linearity of f is quite often automatically assumed, especially if the total cost of producing each bulk P_i is known, in which case

$$\alpha_i = \frac{C_i}{\hat{\boldsymbol{m}}_i(P_i)}$$

where C_i is the total cost of producing and providing the bulk P_i. Our approach allows us to introduce nonlinear cost functions which may be more appropriate in real situations, e.g., in the case of rating electricity where the cost function may contain quadratic terms since transport losses are proportional to the square of the quantity of electricity transported.

Example. Consider an electric power plant serving a large city. Since the demand by the customers varies greatly during a day, the cost of providing a unit of electricity varies according to the quantity $c(t)$ provided at each moment. However, these quantities $c(t)$ may be approximated by their respective averages during the following time intervals which leads to the types P_1, P_2, P_3, P_4:

Type	**Time interval**
P_1	5 a.m.–10 a.m.
P_2	10 a.m–4 p.m.
P_3	4 p.m.–11 p.m.
P_4	11 p.m.–5 a.m.

Table 28.1: Types of services provided by an electric power plant

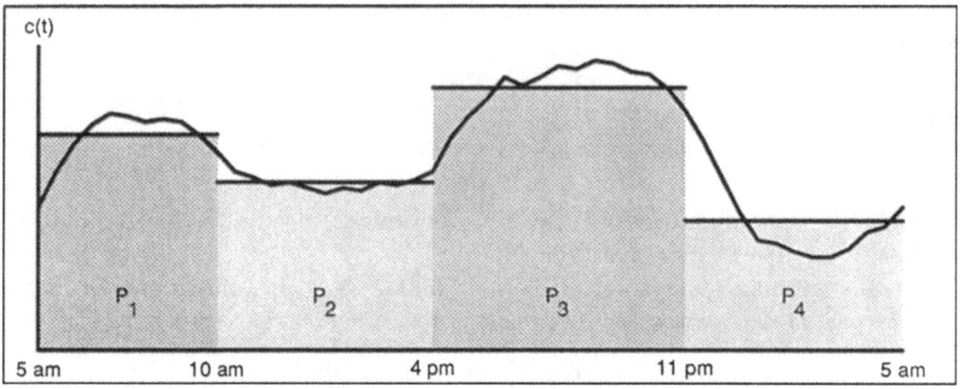

Figure 28.1: Quantities and types of electricity provided

Suppose that a regression analysis delivered a quadratic cost function

$$f(q_1, q_2, q_3, q_4) = \sum_{i=1}^{4} b_{i0} \cdot q_i + \sum_{i=1}^{4} \sum_{j=1}^{4} b_{ij} \cdot q_i \cdot q_j.$$

Then (28.4) yields for $i \in \{1, 2, 3, 4\}$

$$r_i = b_{i0} + \sum_{j=1}^{4} b_{ij} \boldsymbol{m}_j(P_j)$$

where

$$\boldsymbol{m}_j(P_j) = \int_{I_j} c(t)\, \mathrm{d}t,$$

I_j being the time interval corresponding to the type P_j.

29 Plausibility Measures, Possibility Measures, and Normalized Valuations on MV Algebras

29.1 In this section we give a few pointers showing how the measure theoretical part of this book can be put into a more general framework, relating the probability T-measures introduced in Definition 5.3 to other ways of describing uncertainty in environments based on many-valued logics.

On the one hand, probability T-measures defined on a T-tribe $\mathscr{T}$ which contains all the constant fuzzy subsets of the universe X are special cases of *plausibility measures* which are monotone and in a certain way subadditive rather than being T-valuations. Plausibility measures have been introduced by SHAFER [1976]; they were found to be useful for the description of allocations of probability [SHAFER 1979].

On the other hand, $[0,1]^X$, equipped with $\mathbf{S}_\infty$ and $\mathbf{T}_\infty$, is an MV algebra as defined by CHANG [1958]. In many-valued logics, MV algebras play exactly the same role as Boolean algebras in traditional binary logic, and MV algebras are in a one-to-one correspondence with certain approximately finite-dimensional C^*-algebras which are useful in quantum statistical mechanics. Moreover, with the exception of the continuity from the left, our axioms for probability T_∞-measures on $[0,1]^X$ are exactly the same as those for *normalized valuations* [MUNDICI 1991] defined on an MV algebra. All this means that our considerations concerning T_∞-measures have also interpretations outside the game theoretical context.

29.2 We first look at plausibility and possibility measures. Let T be a Borel measurable t-norm, S its corresponding t-conorm, X a nonempty set and $\mathcal{T} \subseteq [0,1]^X$ a T-tribe which contains all the constant fuzzy subsets of X, i.e., which satisfies

$$\alpha X \in \mathcal{T} \qquad \text{for all} \quad \alpha \in [0,1],$$

and which is also a T_0-tribe. Then $\mathcal{T}$ is a T-fuzzy σ-algebra in the sense of [HÖHLE & KLEMENT 1984]. Note that the conditions above are satisfied by each generated T-tribe and for each T_s-tribe with $0 < s \leq \infty$ (Proposition 3.3). For a nongenerated T_0-tribe which is a T_0-fuzzy σ-algebra see Example 2.6 (iv). We recall the definitions of plausibility measures and T-possibility measures from [HÖHLE & KLEMENT 1984]:

A *plausibility measure* $\boldsymbol{m}$ on $\mathcal{T}$ is a function $\boldsymbol{m} : \mathcal{T} \longrightarrow [0,1]$ which is monotone as defined in (5.4), continuous from the left (5.5), satisfies the boundary conditions

$$\boldsymbol{m}(\emptyset) = 0, \qquad \boldsymbol{m}(X) = 1, \tag{29.1}$$

and is subadditive in the sense that for all $A_1, A_2, \ldots, A_n \in \mathcal{T}$ we have

$$\boldsymbol{m}(\mathop{\mathbf{T}_0}_{k=1}^{n} A_k) \leq \sum_{k=1}^{n} \left[(-1)^{k-1} \cdot \sum_{1 \leq j_1 < \ldots < j_k \leq n} \boldsymbol{m}(\mathop{\mathbf{S}_0}_{i=1}^{k} A_{j_i}) \right].$$

A T-*possibility measure* $\boldsymbol{m}$ on $\mathcal{T}$ is a function $\boldsymbol{m} : \mathcal{T} \longrightarrow [0,1]$ which is monotone (5.4), continuous from the left (5.5), satisfies the boundary conditions (29.1), and also satisfies for all $A, B \in \mathcal{T}$

$$\boldsymbol{m}(A\,\mathbf{S}\;B) = \max(\boldsymbol{m}(A), \boldsymbol{m}(B)).$$

Plausibility measures were first introduced in [SHAFER 1976], the concept of possibility goes back to [ZADEH 1978].

29.3 Since the T-tribe $\mathcal{T}$ considered in 29.2 is a distributive lattice with respect to the pointwise natural order (even if the t-norm T and its corresponding t-conorm S are not distributive with respect to each other), it follows from Corollary 2.2 in [HÖHLE & KLEMENT 1984] that each probability T-measure is also a plausibility measure; the same Corollary 2.2 also states that each T-possibility measure is a plausibility measure too. This means that plausibility measures provide a common and rather general framework for both probability T-measures and T-possibility measures. Without going into great detail, we can say that the family of plausibility measures on $\mathcal{T}$ is in a one-to-one correspondence with a specific subclass of regular Borel probability measures on the set of all *realizations* of $\mathcal{T}$ [HÖHLE & KLEMENT 1984, Theorem 3.3]. Moreover, it is possible to characterize the probability T-measures by yet another subclass of this family of regular Borel probability measures on the set of *realizations* of $\mathcal{T}$ [HÖHLE & KLEMENT 1984, Corollary 3.6]. These two results also provide some kind of bridge between plausibility measures and probability T-measures, on the one hand, and classical probabilities, on the other hand, in a way which is quite different from the Representation Theorems 5.8, 6.2 and 7.1 in this book.

29.4 Switching to the second topic of this section, recall the definition of an MV algebra given by CHANG [1958]. An *MV algebra* $\mathfrak{A}$ is a tuple $(A, \oplus, \bullet, ^*, 0, 1)$ such that $(A, \oplus)$ is an abelian monoid and the following properties hold for all $x, y \in A$:

$$\begin{aligned} x \oplus 1 &= 1, \\ x^{**} &= x, \\ 0^* &= 1, \\ (x^* \oplus y)^* \oplus y &= (y^* \oplus x)^* \oplus x, \\ x \bullet y &= (x^* \oplus y^*)^*. \end{aligned}$$

A partial order can be defined on an MV algebra $\mathfrak{A}$ by

$$x \leq y \qquad \Leftrightarrow \qquad (x \bullet y^*) \oplus y = y,$$

turning $\mathfrak{A}$ into a distributive lattice.

A canonical example of an MV algebra is $([0,1], S_\infty, T_\infty, ^*, 0, 1)$, where $x^* = 1-x$. An MV algebra is a Boolean algebra if and only if $\oplus$ is idempotent, i.e., for all $x \in A$ we have $x \oplus x = x$ [CHANG 1958]. In many-valued logic, MV algebras play a similar role as Boolean algebras do in classical logic.

29.5 Obviously, for each nonempty set X, $([0,1]^X, \mathbf{S}_\infty, \mathbf{T}_\infty, ^*, \emptyset, X)$ and, by Proposition 2.7, each T_s-tribe with $0 < s \leq \infty$, is an MV algebra, again with $x^* = 1 - x$. Moreover, CHANG [1958] has shown that each semisimple MV algebra (i.e., an MV algebra in which the intersection of maximal ideals is always zero) is isomorphic to the MV algebra $([0,1]^X, \mathbf{S}_\infty, \mathbf{T}_\infty, ^*, \emptyset, X)$ for some suitable set X.

DI NOLA [1992] extended this representation to the most general case, possibly allowing infinitesimal real values: up to isomorphism, each MV algebra is an algebra of functions from some suitable set X into $[0,1]^*$, where $[0,1]^*$ is some ultrapower of the real unit interval $[0,1]$.

Given a lattice ordered group $(G, +, 0, \vee, \wedge)$ with order unit u (i.e., such that for each $x \in G$ there is an $n \in \mathbb{N}$ such that $x \leq nu$), then $([0,u], \oplus, \bullet, ^*, 0, u)$, is an MV algebra, where

$$\begin{aligned} x \oplus y &= u \wedge (x+y), \\ x^* &= u - x, \\ x \bullet y &= (x^* \oplus y^*)^*. \end{aligned}$$

MUNDICI [1986, Theorem 3.8] proved that the converse is also true, i.e., that for each MV algebra $\mathfrak{A}$ there exists a lattice ordered group G with order unit such that $\mathfrak{A}$ is isomorphic to $[0,u]$ as above; G itself is unique up to isomorphism.

A C^*-algebra is said to be *approximately finite–dimensional* (AF) if it is the inductive limit of an increasing sequence of finite–dimensional C^*-algebras, all with the same unit. AF C^*-algebras are the standard tool for the description of, e.g., spin systems in quantum statistical mechanics. Countable MV algebras are categorically equivalent to certain AF C^*-algebras [MUNDICI 1986, Theorem 3.12].

29.6 Given an MV algebra $(A, \oplus, \bullet, ^*, 0, 1)$, a *normalized valuation* on $\mathfrak{A}$ is a function $\boldsymbol{m} : A \longrightarrow [0,1]$ such that for all $x, y \in A$

$$\boldsymbol{m}(0) = 0, \qquad \boldsymbol{m}(1) = 1,$$

$$\boldsymbol{m}(x \oplus y) + \boldsymbol{m}(x \bullet y) = \boldsymbol{m}(x) + \boldsymbol{m}(y).$$

In the light of the representations mentioned in 29.4, it is clear that each probability T_∞-measure on $[0,1]^X$ is a normalized valuation on $[0,1]^X$. Normalized valuations on MV algebras are in a one–to–one correspondence with invariant measures on the projections of the corresponding AF C^*-algebras [MUNDICI 1991]. As a consequence, it will be interesting to see whether our results can be used in quantum mechanics.

30 Open Problems and Bibliographical Comments

30.1 Summarizing our attempt to present a unified approach to triangular norm-based measures and cooperative games with fuzzy coalitions, and to emphasize the close relationship between these concepts, we have discussed the following problems:

(i) the relationship between the algebraic and topological properties of a t-norm T, on the one hand, and the internal measurability structure of the corresponding T-tribes, on the other hand;

(ii) an integral representation of triangular norm–based measures;

(iii) the geometric form of the range of triangular norm–based vector measures;

(iv) the existence of Aumann–Shapley values for games with fuzzy coalitions defined on not necessarily generated tribes which can be computed via a diagonal formula (and therefore satisfy the diagonal property);

(v) the interconnection between Aumann–Shapley values and equilibria of market games.

The degree of generality of our results was determined by our current knowledge of the mathematical objects we analyzed. So far this knowledge is still quite limited, as is also the case in other fields which are in an early state of development. What we present is an account of research mostly carried out within the past twenty years. Because of the limited technical development in the area, some of the problems mentioned above may not yet have been solved under the most general hypotheses. Some of the technical assumptions were required in several places in spite of the fact that they may have no natural heuristic interpretations. In what follows, we want to discuss some possible improvements to the solutions presented here, as well as ways of relating them to topics which for various reasons were not included in the main body of our work.

30.2 A basic feature of our approach is the use of t-norms as models of rules for forming and dismantling coalitions in a way which is similar to how aggregation and disaggregation of objects are modelled in quantum physics.

In this respect the fundamental triangular norm–based tribes play the role of the proximal setting in which the interactions between players take place. If T_s is a fundamental t-norm, then a T_s-tribe, equipped with the operations T_s and T_0, forms a particular semiring (see Proposition 2.7 and [GOLAN 1992]). In addition, a T_s-tribe has a built–in measurability structure (Theorem 3.2) which is essential for our results. It is not clear whether there are other t-norms T such that the corresponding T-tribes have a similar measurability structure, i.e., are T_∞-tribes. To specify a solution to the following should be interesting:

Problem. *Characterize the family of t-norms T such that T-tribes are necessarily T_∞-tribes.*

The identification of all these t-norms can be of interest in many–valued logic too: it yields a class of models of infinite propositional calculi, incorporating Łukasiewicz infinite propositional calculus, of which T_∞-tribes are indeed models. Possible elements of this family could be ordinal sums of fundamental t-norms (see 1.9) which, together with their corresponding t-conorms, are the only continuous solutions of the functional equation (1.25), as shown by FRANK [1979]. Continuity and equation (1.25) are crucial for the proof of Theorem 3.2. In spite of the absence of any linearity in the sense of (1.25), it might be worthwile to check whether elements of the families $(H_s)_{s\in[0,\infty]}$, $(R_s)_{s\in[-\infty,+\infty]}$ and $(Y_s)_{s\in[0,\infty]}$, as defined in Examples 1.6, 1.7 and 1.8, respectively, are solutions to the problem.

30.3 For T-measures $\boldsymbol{m}$ defined on T-tribes $\mathcal{T}$ which are also T_∞-tribes (i.e., if T is a solution to Problem 30.2) the measurability structure of $\mathcal{T}$ allows us to look for integral representations of $\boldsymbol{m}$ with respect to the restriction $\check{\boldsymbol{m}}$ of $\boldsymbol{m}$ to the σ-algebra $\mathcal{T}^\vee$. In the Representation Theorems 5.8, 6.2 and 7.1 such an integral representation by means of Markov kernels is given for fundamental t-norms T_s and monotone T_s-measures defined on generated T_s-tribes. Monotonicity and the nature of generated tribes are crucial in the proofs of Theorems 5.8 and 7.1.

Problem. *In Theorems 5.8 and/or 7.1, is it possible to drop the hypothesis that the T_s-tribe be generated and/or the T_s-measure be monotone?*

Theorem 6.2 shows that for T_∞-measures none of these conditions is necessary in order to obtain integral representations as in (7.1). For monotone T_s-measures on nongenerated tribes, the existence of such integral representations seems to be a difficult problem; it is equivalent to the fact that the component $\mathring{\boldsymbol{m}}$ in the decomposition (8.8) is identically zero. However, it may not be hopeless since at the moment we do not know any concrete counterexample.

The existence of an integral representation for a T_s-measure without requiring monotonicity is equivalent to the fact that it can be written as the difference of two monotone T_s-measures (Jordan decomposability). If there are t-norms (different to the fundamental t-norms) solving Problem 30.2, it makes sense to ask analogous questions of them.

30.4 The Decomposition Theorem 8.6 provides a decomposition of monotone T_s-measures $\boldsymbol{m}$ on a T_s-tribe $\mathcal{T}$ into the sum of a monotonically irreducible T_s-measure $\mathring{\boldsymbol{m}}$ and a generated T_s-measure. If the T_s-tribe $\mathcal{T}$ is generated, then $\mathring{\boldsymbol{m}}$ equals the zero measure as a consequence of the Third Representation Theorem 7.1. If $s=\infty$, then the Second Representation Theorem 6.2 shows that $\mathring{\boldsymbol{m}}$ always is the zero measure (even if $\mathcal{T}$ is nongenerated). In [MESIAR, to appear] it was shown that, if the underlying set X is countable, then for $s\in\,]0,\infty]$ each T_s-measure $\boldsymbol{m}$ on a T_s-tribe $\mathcal{T}$ is generated, i.e., $\mathring{\boldsymbol{m}}$ equals the zero measure (cf. Remark 8.8). Moreover, in all examples of T_s-measures $\boldsymbol{m}$ on nongenerated T_s-tribes we have checked so far, $\mathring{\boldsymbol{m}}$ always turned out to be the zero measure.

Problem. *Give an example of a T_s-measure $\boldsymbol{m}$ (with $s \neq \infty$) on a nongenerated T_s-tribe $\mathcal{T}$ such that $\mathring{\boldsymbol{m}}$ is different from the zero measure, or prove that $\mathring{\boldsymbol{m}}$ always equals the zero measure whenever $\boldsymbol{m}$ is a T_s-measure on a T_s-tribe $\mathcal{T}$ (generated or not).*

Our Representation Theorems 5.8, 6.2, and 7.2 are based on the Radon-Nikodym Theorem (and thus, implicitly, on the Axiom of Choice). The Decomposition Theorem 8.6 uses these Representation Theorems and relies heavily on Zorn's Lemma and, consequently, is nonconstructive in nature. Therefore, the Decomposition Theorem 8.6 does not provide the necessary tools for constructing an example of the type which is required above.

30.5 One of the main ideas developed is the construction of Aumann–Shapley value on $pFNA_{\mathcal{T}}$ even if the tribe $\mathcal{T}$ is not necessarily generated (see Chapter IV). A basic tool for doing this is a Liapounoff type theorem (Theorem 14.2) which generalizes both the results of LIAPOUNOFF [1940] and DVORETZKY et al. [1951]. The main reason for restricting ourselves in Chapter IV mostly to games based on T_∞-measures is that we do not know whether in Theorem 14.2 the t-norm T_∞ can be replaced by other fundamental t-norms:

Problem. *Does Theorem 14.2 hold for T_s-measures with $s \neq \infty$?*

If a Liapounoff type theorem holds for T_s-measures for some $s \neq \infty$, then it should be possible to expand Theorem 18.4 to games based on such measures.

An argument in favour of the existence of a Liapounoff type theorem for T_s-measures is given by the different technical aspects in various studies of the range of vector measures (see [POPRUGENKO 1928], [CHERNOFF 1951], [MARCUS 1962], [LINDENSTRAUSS 1966]) suggesting that compactness of the range is achievable even if the *linearity* assumption is weakened, somehow compensating this loss by requiring sufficient *continuity*.

30.6 In Proposition 17.2 we show that a game $\boldsymbol{v} = f \circ \boldsymbol{m}$, $\boldsymbol{m}$ being a nonatomic T_∞-measure, is in $pFNA_{\mathcal{T}}$ if and only if f is absolutely continuous. In Theorem 23.6 such a game is shown to belong to $GDIFFE_{\mathcal{T}}$ whenever f has a finite derivative $f'(t)$ at each point t. Moreover, it is easy to prove that the latter is true even if f is continuous and all lateral derivatives $f'(t^+)$, $f'(t^-)$ are finite (without being necessarily equal).

Problem. *Identify the class $\mathcal{G}$ of functions $f : [0,1] \longrightarrow \mathbb{R}$ such that the game $\boldsymbol{v} = f \circ \boldsymbol{m}$ belongs to $GDIFF(E)_{\mathcal{T}}$ whenever $\boldsymbol{m} : \mathcal{T} \longrightarrow \mathbb{R}$ is a probability T_∞-measure. Is the class of games $\boldsymbol{v} = f \circ \boldsymbol{m}$ with $f \in \mathcal{G}$ dense in $GDIFF(E)_{\mathcal{T}}$?*

Such a class of *simple* games $\boldsymbol{v} = f \circ \boldsymbol{m}$ with $f \in \mathcal{G}$ which is dense in $GDIFF(E)_{\mathcal{T}}$ would considerably simplify the computation of the Aumann–Shapley (semi-)value.

An analogous problem can be stated for $GDIFF(E)^{*}_{\mathcal{T}}$.

30.7 The following inclusions between spaces with Aumann–Shapley values are immediate consequences of Theorem 20.4, Proposition 23.2, the considerations in 25.1 and 25.5, and Definition 25.3:

$$pFNA_{\mathcal{T}} \subseteq GDIFFE_{\mathcal{T}} \cap MDIFFE_{\mathcal{T}} \subset GDIFFE^{*}_{\mathcal{T}} \cap MDIFFE^{*}_{\mathcal{T}} \tag{30.1}$$

Problem. *Determine whether the first inclusion in (30.1) is an equality or a strict inclusion.*

Note that the second inclusion in (30.1) is indeed strict: the 2–handed glove market belongs to $GDIFFE^{*}_{\mathcal{T}} \cap MDIFFE^{*}_{\mathcal{T}}$, but not to $GDIFFE_{\mathcal{T}} \cap MDIFFE_{\mathcal{T}}$ (see Example 25.5 and [MERTENS 1988]).

30.8 Aumann–Shapley value operators based on various kinds of differentiability, as presented here, apply to spaces of games with some regularity along the diagonal of the unit cube. However, significant classes of games are not covered, e.g., the *3-handed glove market* [AUMANN & SHAPLEY 1974, p. 201–204].

Problem. *Find a diagonal Aumann–Shapley value on a space which contains* $GDIFFE^{*}_{\mathcal{T}}$ *and the* n*-handed glove markets.*

If it exists, such an extension will require a different type of differential operator since we have already shown that all these games have Dini derivatives (Section 25). For more information about these games see [HART 1980], [TAUMANN 1981] and [MONDERER 1989].

30.9 In Proposition 27.7 we show that the extended market game is well-defined and if the utility function u is sufficiently smooth, then $\hat{\boldsymbol{v}} \in GDIFFE_{\mathcal{T}}$, and the value $\theta\hat{\boldsymbol{v}}$ provides the competitive payoff distribution of a transferable utility competitive equilibrium. This gives a partial answer to a problem raised by AUMANN & SHAPLEY [1974, §41, Open Problem C]. In fact, Aumann & Shapley ask whether the same result can be proven, requiring only that the market game $\boldsymbol{v} = (\hat{\boldsymbol{v}})^{\vee}$ be well-defined (rather than the extended market game $\hat{\boldsymbol{v}}$), at the same time keeping the smoothness hypothesis imposed upon u.

Problem. *Prove or disprove that Proposition 27.7 still holds if we replace the hypothesis that* $\hat{\boldsymbol{v}}$ *be well-defined by the following: there exists an allocation* f *such that*

$$\hat{\boldsymbol{v}}(X) = \int_X u(x, f(x))\, \mathrm{d}\boldsymbol{q}(x). \tag{30.2}$$

Observe that Open Problem C in [AUMANN & SHAPLEY 1974, §41] is quite similar. The main difference is that in our problem the existence of a suitable allocation satisfying (30.2) is required for the extended market game $\hat{\boldsymbol{v}}$ rather than for the market game $\boldsymbol{v}$, and that the σ-additive measure $\boldsymbol{q}$ in our approach is not necessarily is nonatomic. It may be that the comparably rich coalitional structure of the extended market games $\hat{\boldsymbol{v}}$ is the key tool to solve our problem and, implicitly, to give a complete answer to the original problem.

Bibliography

ACZEL, J. (1969) *Lectures on Functional Equations and Their Applications.* Academic Press, New York.

ACZEL, J. and C. ALSINA (1984) Characterizations of some classes of quasilinear functions with applications to triangular norms and synthesis judgements. In R. HENN and D. PALLASCHKE (Eds.) *Contributions to Production Theory, Natural Resources, Economic Indices and Related Topics. Methods of Oper. Res.* **48**: 3–21, Verlagsgruppe Athenäum, Hanstein.

ARSTEIN, Z. (1971) Values of games with denumerably many players. *Int. J. Game Theory* **1**: 27–37.

AUBIN, J.P. (1974a) Coeur et valeur des jeux flous à paiements latéraux. *C.R. Acad. Sci. Paris* **279 A**: 891–894.

AUBIN, J.P. (1974b) Coeur et équilibres des jeux flous sans paiements latéraux. *C.R. Acad. Sci. Paris* **279 A**: 963–966.

AUBIN, J.P. (1981) Locally Lipschitz cooperative games. *J. Math. Economics* **8**: 241–262.

AUBIN, J.P. (1982) *Mathematical Methods of Game and Economic Theory.* North-Holland, Amsterdam.

AUBIN, J.P. (1984) Cooperative fuzzy games: The static and dynamic points of view. *TIMS / Studies in the Management Sciences* **20**: 407–428. Elsevier, Amsterdam.

AUMANN, R.J. and M. PERLES (1965) A variational problem arising in economics. *J. Math. Anal. Appl.* **11**: 488–503.

AUMANN, R.J. and L.S. SHAPLEY (1974) *Values of Non-Atomic Games.* Princeton University Press, Princeton.

BAUER, H. (1972) *Probability Theory and Elements of Measure Theory.* Holt, Rinehart & Winston, New York.

BILLERA, L.J., D.C. HEATH and R.J. RAANAN (1978) Internal telephone billing rates — A novel application of non-atomic game theory. *Operations Research* **26**: 956–965.

BILLOT, A. (1990) Fuzzy convexity and peripherial core of an exchange economy represented as a fuzzy game. In J. KACPRZYK and M. FEDRIZZI (Eds.) *Multipersons Decision Making Models using Fuzzy Sets and Possibility Theory.* Kluwer, Dordrecht.

BLACKWELL, D. (1951) The range of certain vector integrals. *Proc. Amer. Math. Soc.* **2**: 390–395.

BONDAREVA, O.N. (1963) Some applications of linear programming methods to the theory of cooperative games (in Russian). *Prob. Kibernet.* **10**: 119–139.

BUTNARIU, D. (1983a) Additive fuzzy measures and integrals I. *J. Math. Anal. Appl.* **93**: 436–452.

BUTNARIU, D. (1983b) Decompositions and ranges of additive fuzzy measures. *Fuzzy Sets and Systems* **10**: 135–155.

BUTNARIU, D. (1985) Non-atomic fuzzy measures and games. *Fuzzy Sets and Systems* **17**: 39–52.

BUTNARIU, D. (1986) Fuzzy measurability and integrability. *J. Math. Anal. Appl.* **117**: 385–410.

BUTNARIU, D. (1987a) Additive fuzzy measures and integrals III. *J. Math. Anal. Appl.* **125**: 288–303.

BUTNARIU, D. (1987b) Values and cores for fuzzy games with infinitely many players. *Int. J. Game Theory* **16**: 43–68.

BUTNARIU, D. (1989) A fuzzy game theoretical approach of the rate problem for services in bulk. In J.L. VERDEGAY and M. DELGADO (Eds.) *The Interface between Artificial Intelligence and Operations Research.* pp. 35–50. TÜV, Basel.

BUTNARIU, D. and Y. CENSOR (1991) On a class of bargaining schemes for points in the core of a cooperative n-person game. Technical Report 280, Research Center for Advanced Study, Univ. of Texas at Arlington, Arlington.

BUTNARIU, D. and E.P. KLEMENT (1991) Triangular norm-based measures and their Markov kernel representation. *J. Math. Anal. Appl.* **162**: 111–143.

CHABRILLAC, Y. and J.-P. CROUZEIX (1987) Continuity and differentiability properties of monotone real functions of several real variables. *Mathematical Programming Study* **30**: 1–16.

CHANG, C.C. (1958) Algebraic analysis of many valued logics. *Trans. Amer. Math. Soc.* **88**: 467–490.

CHERNOFF, H. (1951) An extension of a result of Liapounoff on the range of a vector measure. *Proc. Amer. Math. Soc.* **2**: 720–726.

CLIMESCU, A.C. (1946) Sur l'équation fonctionelle de l'associativité. *Bull. École Polytechnique Iassy* **1**: 1–16.

CROUZEIX, J.-P. (1982) A review of continuity and differentiability properties of quasiconvex functions on $\mathbb{R}^n$. In J.P. AUBIN and R.B. VINTER (Eds.) *Convex Analysis and Optimization.* pp. 18–34, Pitman, Boston.

DI NOLA, A. (1992) Representation and reticulation by quotients of MV algebras. *Ricerche di Matematica Napoli* **40**.

DUNFORD, N. and J.T. SCHWARTZ (1958) *Linear Operators.* Interscience, New York.

DVORETZKY, A., A. WALD and J. WOLFOWITZ (1950) Elimination of randomization in certain problems of statistics and game theory. *Proc. Nat. Acad. Sc. USA* **36**: 256–259.

DVORETZKY, A., A. WALD and J. WOLFOWITZ (1951) Relations among certain ranges of vector measures. *Pac. J. Math.* **1**: 59–74.

FRANK, M.J. (1979) On the simoultaneous associativity of $F(x,y)$ and $x+y-F(x,y)$. *Aequationes Math.* **19**: 194–226.

FUSTIER, B. (1986) The fuzzy demand. In C. PONSARD and B. FUSTIER (Eds.) *Fuzzy Economics and Spatial Analysis.* Collection de l'Institut de Mathématiques Economiques **32**: 29–45, Université de Bourgogne, Dijon.

GOLAN, J.S. (1992) *Theory of semirings with Applications in Mathematics and Theoretical Computer Science.* Longmann, New York.

HALMOS, P.R. (1947) On the set of values of a finite measure. *Bull. Amer. Math. Soc.* **53**: 138–141.

HALMOS, P.R. (1948) The range of a vector measure. *Bull. Amer. Math. Soc.* **54**: 416–421.

HALMOS, P.R. (1974) *Measure Theory.* Springer, Berlin.

HAMACHER, H. (1978) *Über logische Aggregationen nicht-binär explizierter Entscheidungskriterien.* Rita G. Fischer Verlag, Frankfurt.

HART, S. (1980) Measure based values of market games. *Math. Oper. Res.* **2**: 197–221.

HAWKINS, T. (1970) *Lebesgue's Theory of Integration: Its Origins and Development.* University of Wisconsin Press, Madison.

HILDENBRAND, W. (1974) *Core and Equilibria in a Large Economy,* Princeton University Press, Princeton

HÖHLE, U. (1976) Maße auf unscharfen Mengen. *Z. Wahrsch. Verw. Gebiete* **36**: 179–188.

HÖHLE, U. and E.P. KLEMENT (1984) Plausibility measures - A general framework for possibility and fuzzy probability measures. In H.J. SKALA, S. TERMINI and E. TRILLAS (Eds.) *Aspects of Vagueness.* pp. 31–50, Reidel, Dordrecht.

HOLMES, R.B. (1975) *Geometric Functional Analysis and Applications.* Springer, New York.

KANNAI, Y. (1966) Values of games with a continuum of players, *Israel J. Math.* **4**: 54–58.

KANNAI, Y. (1969) Countably additive measures in cores of games. *J. Math. Anal. Appl.* **27**: 227–240.

KANTOROVITSCH, L.W. and G.P. AKILOV (1964) *Functional Analysis.* Pergamon Press, Oxford.

KLEMENT, E.P. (1980a) Fuzzy σ–algebras and fuzzy measurable functions. *Fuzzy Sets and Systems* **4**: 83–93.

KLEMENT, E.P. (1980b) Characterization of finite fuzzy measures using Markoff-kernels. *J. Math. Anal. Appl.* **75**: 330–339.

KLEMENT, E.P. (1982a) Construction of fuzzy σ–algebras using triangular norms. *J. Math. Anal. Appl.* **85**: 543–565.

KLEMENT, E.P. (1982b) Characterization of fuzzy measures constructed by means of triangular norms. *J. Math. Anal. Appl.* **86**: 345–358.

KLEMENT, E.P., R. LOWEN and W. SCHWYHLA (1981) Fuzzy probability measures. *Fuzzy Sets and Systems* **5**: 21–30.

KLIR, G.J. and T.A. FOLGER (1988) *Fuzzy Sets, Uncertainity and Information.* Prentice Hall, Englewood Cliffs.

KOLMOGOROFF, A.N. (1933) *Grundbegriffe der Wahrscheinlichkeitsrechnung.* Springer, Berlin.

LIAPOUNOFF, A.A. (1940) Sur les fonctions-vecteurs completement additives (in Russian). *Bull. Acad. Sci. URSS Ser. Math.* **4**: 465–478.

LINDENSTRAUSS, J. (1966) A short proof of Liapounoff's convexity theorem. *J. Math. Mech.* **15**: 971–973.

MARCUS, S. (1962) Atomic measures and their Darboux property. *Rev. Roum. Math. Pures et Appl.* **2**: 327-333.

MENGER, K. (1942) Statistical metrics. *Proc. Nat. Acad. Sci. U.S.A.* **28**: 535–537.

MENGER, K. (1951a) Probabilistic theories of relations. *Proc. Nat. Acad. Sci. U.S.A.* **37**: 178–180.

MENGER, K. (1951b) Probabilistic geometry. *Proc. Nat. Acad. Sci. U.S.A.* **37**: 226–229.

MERTENS, J.-F. (1980) Values and derivatives. *Math. Oper. Res.* **5**: 523–552.

MERTENS, J.-F. (1988) The Shapley value in the non-differentiable case. *Int. J. Game Theory* **17**: 1–65.

MILNOR, J.W. and L.S. SHAPLEY (1961) Values of large games II: Oceanic games. *RM-2649, The Rand Cooperation*, Santa Monica, California.

MESIAR, R. Fundamental triangular norm-based tribes and measures. *J. Math. Anal. Appl.*, to appear.

MONDERER, D. (1986) Measure-based values of nonatomic games, *Math. Oper. Res.* **11**: 321–335.

MONDERER, D. (1989) Asymptotic measure-based values of non-atomic games, *Math. Oper. Res.* **14**: 737–745.

MONDERER, D. and A. NEYMANN (1988) Values of smooth non-atomic games: The method of multilinear approximation. In A.E. ROTH (Ed.) *The Shapley Value: Essays in Honor of Lloyd S. Shapley.* pp. 217–234, Cambridge University Press, Cambridge.

MONDERER, D. and W.H. RUCKLE (1990) On the symmetry axiom for values of non-atomic games. *Int. J. Math. and Math. Sciences* **13**: 165–170

MUNDICI, D. (1986) Interpretation of AF C^*-algebras in Łukasiewicz Sentential Calculus. *J. Funct. Anal.* **65**: 15–63.

MUNDICI, D. (1991) Uncertainty measures in MV algebras, and states of AF C^*-algebras. In E.P. KLEMENT and S. WEBER (Eds.) *Abstracts of the 13th Linz Seminar on Fuzzy Set Theory: Uncertainty Measures.* pp. 42–43. Johannes Kepler Universität, Linz.

NATANSON, I.P. (1969) *Theorie der Funktionen einer reellen Veränderlichen.* Akademie-Verlag, Berlin.

NEUMANN, J. (VON) (1928) Zur Theorie der Gesellschaftsspiele. *Math. Ann.* **100**: 295–320.

NEUMANN, J. (VON) and O. MORGENSTERN (1944) *Theory of Games and Economic Behavior.* Princeton University Press, Princeton.

OWEN, G. (1972) Multilinear extensions of games. *Manag. Sci.* **18**: 64–79. See also the reedited version in A.E. ROTH (Ed., 1988) *The Shapley Value: Essays in*

Honor of Lloyd S. Shapley. pp. 139–151, Cambridge University Press, Cambridge.

PARKER, J.M. (1991) The sigma–core of a cooperative game. *Manuscripta Math.* **70**: 247–253.

PARTHASARATHY, K.R. (1978) *Introduction to Probability and Measures.* McMillan, New Delhi.

PONSARD, C. (1986a) A theory of spatial general equilibrium in a fuzzy economy. In C. PONSARD and B. FUSTIER (Eds.) *Fuzzy Economics and Spatial Analysis.* Collection de l'Institut de Mathématiques Economiques **32**: 1–27, Université de Bourgogne, Dijon.

PONSARD, C. (1986b) Spatial oligopoly as fuzzy game. In C. PONSARD and B. FUSTIER (Eds.) *Fuzzy Economics and Spatial Analysis.* Collection de l'Institut de Mathématiques Economiques **32**: 57–67, Université de Bourgogne, Dijon.

POPRUGENKO, G. (1928) Sur la propriété de Darboux des fonctions continues d'ensemble. *Fundam. Math.* **12**: 254–268.

ROCKAFELLAR, R.T. (1970) *Convex Analysis.* Princeton University Press, Princeton.

RUCKLE, W.H. (1982) An extension of the Aumann–Shapley value concept to functions on arbitrary Banach spaces. *Int. J. Game Theory* **11**: 105–116.

RUCKLE, W.H. (1987) The linearizing projection, local theories. Technical Report 534, Department of Mathematical Sciences, Clemson University, Clemson.

SCHMEIDLER, D. (1969) Competitive equilibria in markets with a continuum of traders and incomplete preferences. *Econometrica* **37**: 578–585.

SCHMEIDLER, D. (1972) Cores of exact games I. *J. Math. Anal. Appl.* **40**: 214–225.

SCHMIDT, K.D. (1982) A general Jordan decomposition. *Arch. Math.* **38**: 556–564.

SCHMIDT, K.D. (1985) A common abstraction of Boolean rings and lattice ordered groups. *Comp. Math.* **54**: 51–62.

SCHMIDT, K.D. (1986) Decompositions of vector measures in Riesz spaces and Banach lattices. *Proc. Edinburgh Math. Soc.* **29**: 23–39.

SCHMIDT, K.D. (1989) *Jordan Decompositions of Generalized Vector Measures.* Longman, Essex.

SCHWEIZER, B. and A. SKLAR (1960) Statistical metric spaces. *Pac. J. Math.* **10**: 313–334.

SCHWEIZER, B. and A. SKLAR (1961) Associative functions and statistical triangle inequalities. *Publ. Math. Debrecen* **8**: 169–186.

SCHWEIZER, B. and A. SKLAR (1963) Associative functions and abstract semigroups. *Publ. Math. Debrecen* **10**: 69–81.

SCHWEIZER, B. and A. SKLAR (1983) *Probabilistic Metric Spaces.* North-Holland, New York.

SHAFER, G. (1976) *A Mathematical Theory of Evidence.* Princeton University Press, Princeton.

SHAFER, G. (1979) Allocations of probability. *Annals of Prob.* **7**: 827–839.

SHAPLEY, L.A. (1953) A value for n–person games. In H. KUHN and A. W.

TUCKER (Eds.) *Contributions to the Theory of Games, II.* pp. 307–317, Princeton University Press, Princeton.

SHAPLEY, L.S. (1967) On balanced sets and cores. *Nav. Res. Log. Q.* **14**: 453–460.

SHAPLEY, L.S. (1971) Cores of convex games. *Int. J. Game Theory* **1**: 12-26.

SHAPLEY, L.S. (1973) On balanced games without side payments. In T.C. HU and S.M. ROBINSON (Eds.) *Mathematical Programming.* pp. 261–290, Academic Press, New York.

SOBCZYK, A. and P.C. HAMMER (1944) The range of an additive set function. *Duke Math. J.* **11**: 847–851.

SUGENO, M. (1974) Theory of fuzzy integrals and its applications. Ph.D.Thesis. Tokyo Inst. of Technology, Tokyo.

TAUMANN, Y. (1981) Values of a class of non–differentiable market games. *Int. J. Game Theory* **10**: 155–162.

WALD, A. (1943) On statistical generalizations of metric systems. *Proc. Nat. Acad. Sc. U.S.A.* **29**: 196–197.

WEBER, S. (1984a) $\perp$-decomposable measures and integrals for Archimedean t-conorms $\perp$. *J. Math. Anal. Appl.* **101**: 114–138.

WEBER, S. (1984b) Measures of fuzzy sets and measures of fuzziness. *Fuzzy Sets and Systems* **13**: 247–271.

WYLER, O. (1966) Clans. *Comp. Math.* **17**: 172–189.

YAGER, R.R. (1980) On a general class of fuzzy connectives. *Fuzzy Sets and Systems* **4**: 235–242.

ZADEH, L.A. (1965) Fuzzy sets. *Inform. Control* **8**: 338–353.

ZADEH, L.A. (1968) Probability measures of fuzzy events. *J. Math. Anal. Appl.* **23**: 421–427.

ZADEH, L.A. (1978) Fuzzy sets as a basis for a theory of possibility. *Fuzzy Sets and Systems* **1**: 3–28.

ZIMMERMANN, H.J. (1991) *Fuzzy Set Theory – and Its Applications.* Kluwer, Dordrecht.

List of Symbols

Index

F

G

H

I

J

L

U

V

W

THEORY AND DECISION LIBRARY

SERIES C: GAME THEORY, MATHEMATICAL PROGRAMMING AND OPERATIONS RESEARCH

Editor: S.H. Tijs, *University of Nijmegen, The Netherlands*

1. B.R. Munier and M.F. Shakun (eds.): *Compromise, Negotiation and Group Decision.* 1988 ISBN 90-277-2625-6
2. R. Selten: *Models of Strategic Rationality.* 1988 ISBN 90-277-2663-9
3. T. Driessen: *Cooperative Games, Solutions and Applications.* 1988 ISBN 90-277-2729-5
4. P.P. Wakker: *Additive Representations of Preferences.* A New Foundation of Decision Analysis. 1989 ISBN 0-7923-0050-5
5. A. Rapoport: *Experimental Studies of Interactive Decisions.* 1990 ISBN 0-7923-0685-6
6. K.G. Ramamurthy: *Coherent Structures and Simple Games.* 1990 ISBN 0-7923-0869-7
7. T.E.S. Raghavan, T.S. Ferguson, T. Parthasarathy and O.J. Vrieze (eds.): *Stochastic Games and Related Topics.* In Honor of Professor L.S. Shapley. 1991 ISBN 0-7923-1016-0
8. J. Abdou and H. Keiding: *Effectivity Functions in Social Choice.* 1991 ISBN 0-7923-1147-7
9. H.J.M. Peters: *Axiomatic Bargaining Game Theory.* 1992 ISBN 0-7923-1873-0
10. D. Butnariu and E.P. Klement: *Triangular Norm-Based Measures and Games with Fuzzy Coalitions.* 1993 ISBN 0-7923-2369-6

KLUWER ACADEMIC PUBLISHERS – DORDRECHT / BOSTON / LONDON

GPSR Compliance

The European Union's (EU) General Product Safety Regulation (GPSR) is a set of rules that requires consumer products to be safe and our obligations to ensure this.

If you have any concerns about our products, you can contact us on ProductSafety@springernature.com

In case Publisher is established outside the EU, the EU authorized representative is:

Springer Nature Customer Service Center GmbH
Europaplatz 3
69115 Heidelberg, Germany

Batch number: 10823959

Printed by Printforce, the Netherlands